AF251311

Algorithms and Computation
in Mathematics · Volume 7

Editors

E. Becker   M. Bronstein   H. Cohen
D. Eisenbud   R. Gilman

Springer

Berlin
Heidelberg
New York
Barcelona
Hong Kong
London
Milan
Paris
Singapore
Tokyo

Peter Bürgisser

Completeness and Reduction in Algebraic Complexity Theory

With 16 Figures

Springer

Peter Bürgisser
Fachbereich 17 · Mathematik-Informatik
Universität-Gesamthochschule Paderborn
Warburger Strasse 100
33095 Paderborn, Germany
e-mail: pbuerg@math.uni-paderborn.de

Library of Congress Cataloging-in-Publication Data

Bürgisser, Peter, 1962–. Completeness and reduction in algebraic complexity theory / Peter Bürgisser. p. cm. – (Algorithms and computation in mathematics, ISSN 1431-1550; v. 7)
Includes bibliographical references and index.
ISBN 3540667520 (alk. paper)
1. Computational complexity. I. Title. II. Series   QA267.7.B88 2000 511.3–dc21   00-029647

Mathematics Subject Classification (2000): 68Q05, 68Q15, 68Q25, 68Q40, 15A15, 22E70, 33C25, 03D15, 03D25, 05Cxx, 82B20

ISSN 1431-1550

ISBN 3-540-66752-0 Springer-Verlag Berlin Heidelberg New York

Springer-Verlag is a company in the BertelsmannSpringer publishing group.
© Springer-Verlag Berlin Heidelberg 2000
Printed in Germany

Cover design: MetaDesign plus GmbH, Berlin
Cover production; *design & production* GmbH, Heidelberg
Typeset in LaTeX by the author

Printed on acid-free paper   SPIN 10743448   46/3143Ko – 5 4 3 2 1 0

Dedicated to

Brigitte and Ladina

Preface

One of the most important and successful theories in computational complexity is that of NP-completeness. This discrete theory is based on the Turing machine model and achieves a classification of discrete computational problems according to their algorithmic difficulty. Turing machines formalize algorithms which operate on finite strings of symbols over a finite alphabet. By contrast, in algebraic models of computation, the basic computational step is an arithmetic operation (or comparison) of elements of a fixed field, for instance of real numbers. Hereby one assumes exact arithmetic. In 1989, Blum, Shub, and Smale [12] combined existing algebraic models of computation with the concept of uniformity and developed a theory of NP-completeness over the reals (BSS-model). Their paper created a renewed interest in the field of algebraic complexity and initiated new research directions. The ultimate goal of the BSS-model (and its future extensions) is to unite classical discrete complexity theory with numerical analysis and thus to provide a deeper foundation of scientific computation (cf. [11, 101]).

Already ten years before the BSS-paper, Valiant [107, 110] had proposed an analogue of the theory of NP-completeness in an entirely algebraic framework, in connection with his famous hardness result for the permanent [108]. While the part of his theory based on the Turing approach (#P-completeness) is now standard and well-known among the theoretical computer science community, his algebraic completeness result for the permanents received much less attention. The first account of Valiant's algebraic theory with elaborated proofs was von zur Gathen's survey [41]. A more recent treatment, with different proofs, can be found in the last chapter of the book [21] by Bürgisser et al.

In this research monograph, we further develop Valiant's approach, and clarify its connections both to the discrete and to the BSS-model. We think this adds considerably to our understanding of the concept of completeness in algebraic models of computation.

This book is the author's Habilitationsschrift in mathematics at the University of Zürich. It is organized as follows. The introduction overviews the three known theories of NP-completeness and explains our main results in an informal way. After a detailed treatment of Valiant's model in Chap. 2, we proceed in Chap. 3 by showing that the generating functions of various NP (or #P) complete graph properties are complete in Valiant's sense. The proofs are mainly based on graph theoretical constructions.

In Chap. 4, we relate Valiant's model to the classical discrete theory. Unexpectedly, parallel complexity classes enter. We rely on techniques from algebraic geometry and number theory, and our main result hinges on the generalized Riemann hypothesis.

The next chapter is devoted to investigations in the spirit of structural complexity. We prove that any countable poset can embedded in the poset of p-degrees, if Valiant's hypothesis is true. A striking result here is the discovery of a specific family of polynomials (cut enumerators), which is neither complete nor p-computable, provided the polynomial hierarchy does not collapse.

In Chap. 6 we deviate a little from our main line of investigation. We develop a fast algorithm to evaluate irreducible rational matrix representations of complex general linear groups with respect to a symmetry adapted basis (Gelfand-Tsetlin basis). The connection to our main topic then becomes clear in the next chapter.

Immanants are matrix functions defined in terms of characters of the symmetric group, which generalize the permanent and determinant. For a deeper understanding of the amazingly different complexity behaviour of determinants and permanents, they are a natural object to study. Moreover, they give also an indication of the complexity to evaluate single entries of invariant matrices of general linear groups. Our algorithm of Chap. 6 yields upper complexity bounds for immanants, which improve previous bounds due to Hartmann [47] and Barvinok [5]. The main efforts of Chap. 7 consist of proofs that the evaluation of immanants corresponding to certain hook diagrams or rectangular diagrams are complete in Valiant's sense.

Finally, Chap. 8 contains some separation results and establishes a connection between Valiant's and the Blum-Shub-Smale model. We discuss possible directions for a deeper understanding of this connection.

Seventeen conjectures and open problems, distributed throughout the text, suggest further research. Some of them have resisted serious attempts for solution by the author.

Acknowledgments

I owe special thanks to Michael Clausen for his encouragement and feedback throughout this project. I am indebted to Amin Shokrollahi for pointing out to me the paper [5] by Barvinok, and I would like to thank Luis Miguel Pardo for indicating that Koiran's method [67] can be improved using the results of [68, 69]. I thank Lenore Blum, Felipe Cucker, Maurice Rojas, and Steve Smale for their interest and numerous discussions about my work.

I am grateful to the Institute for Applied Mathematics of the University of Zürich for making this research possible. In particular, I would like to thank Andrew Barbour and Peter Gabriel for their support. I am greatly indebted to Steve Smale for inviting me to the Liu Bie Ju Centre for Mathematical Sciences at the City University of Hong Kong, where this work was completed.

I would like to thank the staff at Springer-Verlag Heidelberg for editorial advise and help. I am grateful to Fred and Lilly Hämmerli for their help in the final stage of this project. Finally, I wish to thank my wife Brigitte for her support, patience, and understanding, and our daughter Ladina for her cheerful presence.

Zürich, August 1998 *Peter Bürgisser*

Contents

1

Introduction

We start with a brief description of the classical discrete theory of NP-completeness, give an overview of its generalization to the Blum-Shub-Smale model, and present the main features of Valiant's algebraic model. Then we outline the organization of the book and present some of its highlights in an informal way.

1.1 Classical Model

Computational complexity theory provides a framework for understanding the cost required to solve algorithmic problems. Nowadays, the dominant approach to this theory is based on the computational model of the Turing machine. This provides a formalization of algorithms that operate on finite strings from a finite alphabet. One of the most important and successful concepts developed in computational complexity is that of NP-completeness, originating in the work of Cook [25], Karp [60], and Levin [74]. We are going to describe briefly the main features of this theory. (For details see Garey and Johnson [38] or Papadimitriou [87].)

A language $A \subseteq \{0,1\}^* := \cup_{n \geq 0}\{0,1\}^n$ is a set of finite strings over the alphabet $\{0,1\}$. The length of a string x is denoted by $|x|$. The focus is on decision problems, which are formalized by languages. The complexity class P of polynomial time decidable languages consists of all languages A such that membership of $x \in \{0,1\}^n$ to A can be decided by a deterministic Turing machine in a number of steps, which is bounded by a polynomial in n (p-bounded in n). This class serves as a rough idealization for the problems that are computationally tractable. Another, more subtle class NP is considered, that contains all problems for which a solution can be verified in a polynomial number of steps. Formally, this class can be defined as follows. Let $R \subseteq \{0,1\}^* \times \{0,1\}^*$ be a relation between strings, which is decidable in polynomial time, and balanced in the following sense: there exists a p-bounded function $t: \mathbb{N} \to \mathbb{N}$ satisfying $|y| \leq t(|x|)$ for all $(x,y) \in R$. The language $A = \{x \in \{0,1\}^* \mid \exists y \ (x,y) \in R\}$ is in the class NP, and all languages in NP are obtained this way. Hereby, the string y is interpreted as a (short) witness for the membership of x to A.

A fundamental conjecture, *Cook's hypothesis*, claims that $P \neq NP$. This is undoubtedly the most famous open problem of theoretical computer science.

Although there is a lot of practial evidence towards it, a proof is far out of sight.

Languages A and B are compared by means of reductions. A is said to be polynomial time (many one) reducible to B iff there exists a string function $\rho\colon \{0,1\}^* \to \{0,1\}^*$, which is computable in polynomial time on a Turing machine, such that $A = \rho^{-1}(B)$. A language in NP is called NP-*complete* iff every problem in NP can be polynomial time reduced to it. Usually, a proof that a decision problem is NP-complete, is taken as evidence of intractability. Indeed, such a problem is not in P iff Cook's hypothesis is true.

The importance of the P-NP-theory lies in the fact that countless problems of great practical significance in logic, number theory, and combinatorial optimization have been identified as NP-complete ones. Basic NP-complete problems are for instance the satisfiability problem for Boolean formulas, or the problem to decide whether a given graph has a Hamilton cycle. (Of course, these problems have to be encoded as languages.)

The counting complexity class #P introduced by Valiant [108] is relevant for our purposes. Any problem in NP is defined by a balanced, polynomial time decidable string relation R as above. The associated counting problem is the following: given x, how many y are there such that $(x,y) \in R$? So we ask to compute the function $\phi\colon \{0,1\}^* \to \mathbb{N}$ given by

$$\phi(x) := \#R_x := \#\{y \mid (x,y) \in R\} \ .$$

The class #P is defined as the set of all such functions ϕ. A reduction from the counting problem of R to the counting problem of a string relation S is given by two polynomial time computable functions $\rho\colon \{0,1\}^* \to \{0,1\}^*$ and $\sigma\colon \mathbb{N} \to \mathbb{N}$ (natural numbers encoded in binary) such that $\#R_x = \sigma(\#S_{\rho(x)})$ for all x. This reduction is called parsimonious iff σ is the identity. A counting function in #P is called #P-*complete* iff every counting function in #P can be reduced to it. For instance, the problem to count the number of Hamilton cycles in a given graph can be proven to be #P-complete. It is an open problem whether the counting problem associated with an NP-complete problem is always #P-complete.

Valiant [108] was able to prove that the problem to count all perfect matchings of a given bipartite graph is #P-complete. This is particularly interesting, since the corresponding decision problem, also known as the marriage problem, is solvable in polynomial time. The number of perfect matchings of a bipartite graph G can be expressed as the *permanent*

$$\mathrm{per}(A) := \sum_{\pi \in S_n} A_{1,\pi(1)} \cdots A_{n,\pi(n)}$$

of the adjacency matrix A of G. So we have the following important result.

Theorem 1.1 (Valiant) *The problem to evaluate the permanent of a $0,1$-matrix is #P-complete.*

If $\phi\colon \{0,1\}^* \to \mathbb{N}$ is a counting function as above, we may just ask for the parity of $\phi(x)$. The complexity class *parity polynomial time* $\oplus$P, introduced by Papadimitriou and Zachos [89], comprises all such decision problems. One can prove that the problem to find the parity of the number of Hamilton cycles in a given graph is $\oplus$P-complete with respect to polynomial time reductions. Remarkably, one can compute the parity of the number of perfect matchings of a bipartite graph in polynomial time, since this amounts to evaluating a determinant modulo 2. (Permanent and determinant coincide in characteristic two.) This also implies that there is no parsimonious reduction from the problem to count Hamilton cycles to the problem to count the number of perfect matchings.

1.2 Blum-Shub-Smale Model

Algebraic complexity theory is the study of the intrinsic algorithmic difficulty of algebraic and numeric problems. It is based on models of computation that are adapted to the problems under investigation: straight-line programs and computation trees. The basic computational step is an arithmetic operation (or comparison) of elements of an algebraic structure, for instance of a finite field, or the field of real or complex numbers. In the latter cases, an important idealization is the assumption of exact arithmetic. The BSS-model, introduced by Blum, Shub, and Smale [12], adds a new feature to this, by combining the algebraic models of computation with the concept of *uniformity*, a condition sine qua non in the classical theory of computation. (Before, uniformity was not studied systematically in algebraic complexity, mainly because it is unknown how to exploit the uniformity condition for lower bound proofs.) Instead of considering for each dimension n a separate algorithm solving problems of that dimension (for instance, evaluating the permanent of an n by n matrix), we clearly want to have one uniform algorithm capable of solving problems of any dimension. This aspect is incorporated in the definition of the BSS-machine over a field k with the introduction of shift nodes that are used to access the contents of registers. We do not attempt to give a formal definition of such machines here; instead, the reader is referred to the recent textbook [11] for more details. Such a machine is a finite object except that it comes with a finite number of constants in k. (For instance in the case $k = \mathbb{R}$, these constants might not have a finite description.) It is important that BSS-machines over the field $\mathbb{F}_2$ with two elements are equivalent to Turing machines. This implies that the classical theory of complexity and computation is a special case of the more general theory to be developed.

Let $k^\infty := \cup_{n \geq 0} k^n$ be the set of finite sequences over k. A BSS-machine M over k takes as inputs elements x of k^∞ and computes an output $y \in k^\infty$ if the computation stops. The number of steps spent during this computation measures the number of arithmetic operations and comparisons of elements of k, as well as the number of shift operations used to address registers. We mention that if we restrict ourselves to inputs $x \in k^n$ of a fixed length n, then the BSS-machines become equivalent to the model of computation trees.

Over a fixed field k, one can now define the complexity classes P_k and NP_k in complete analogy with the classical situation. For instance, the class P_k consists of all decision problems A, defined as a subset of k^∞, such that membership of $x \in k^\infty$ to A can be decided by a BSS-machine over k in a p-bounded number of steps in the length n of an input $x \in k^n$. We have a notion of polynomial time reduction between decision problems, by which we define NP-completeness over k.

Are there natural NP-complete problems in this setting? Consider the following decision version of the fundamental problem to find zeros of systems of polynomial equations, called *Hilbert Nullstellensatz* problem (HN):

> Given a system of polynomials $f_1, \ldots, f_r$ in n variables over $\mathbb{C}$,
> decide whether the f_i have a common complex zero.

The length of an input to this problem is the number of coefficients of the polynomials f_i. Note that (HN) is clearly in $NP_\mathbb{C}$, since a solution can be easily verified by plugging in into the polynomials.

Theorem 1.2 (Blum, Shub, Smale) *The Hilbert Nullstellensatz problem is* NP-*complete over the complex numbers.*

We remark that there is an analogous completeness result over the reals. Over the field $\mathbb{F}_2$, a corresponding result easily implies the completeness of the satisfiability problem for Boolean formulas.

In this context, there is the fundamental conjecture that $P_k \neq NP_k$, which we will call the *BSS-hypothesis*. Note that it might depend on the underlying field k. The BSS-hypothesis over $\mathbb{C}$ expresses that there is no polynomial time algebraic algorithm (formalized by BSS-machines), that solves the Hilbert Nullstellensatz problem. Smale [102] considers the question of whether $P \neq NP$ (over $\mathbb{F}_2$ or $\mathbb{C}$) as one of the most important open problems of mathematics, at position three right after Riemann's hypothesis and the Poincaré conjecture!

The ultimate goal of the effort started with the introduction of the BSS-model is to combine ideas developed in theoretical computer science with numerical analysis in order to create a deeper foundation of the latter. The assumption of infinite precision arithmetic in this model is of course very idealized. We remark that there are ongoing efforts towards a more realistic model of computation over the reals which takes into account the conditioning of inputs as well as round-off errors (see the survey [101]).

1.3   Valiant's Model

Already ten years before the BSS-paper, Valiant [107, 110] had proposed an analogue of the theory of NP-completeness in an entirely algebraic framework, in connection with his famous hardness result for the permanent [108] (see also [111]). The goal of this research monograph is to further develop Valiant's proposal, and to clarify its connections both to the classical and to the BSS-model.

Let us first outline the main features of Valiant's model. (A detailed presentation can be found in Chap. 2.) In contrast to the theories explained before, who deal with decision problems, we study here very basic computational problems: the evaluation of multivariate polynomials.

A p-family over a fixed field k is a sequence $f = (f_n)$ of multivariate polynomials such that the number of variables as well as the degree of f_n are p-bounded functions of n. The algorithmic problem to study is the evaluation of these polynomials. An interesting example of a p-family is the permanent family $\mathrm{PER} = (\mathrm{PER}_n)$, where PER_n is the permanent of an n by n matrix with independent indeterminate entries.

Let $L(f_n)$ denote the total complexity of f_n, that is, the minimum number of arithmetic operations $+, -, *$ sufficient to compute f_n from the variables X_i and constants in k by a straight-line program Γ_n (for a formal definition see Sect. 2.1). A p-family (f_n) is called p-*computable* iff $L(f_n)$ is a p-bounded function of n. The p-computable families constitute the complexity class VP. (V is an acronym for Valiant.)

We remark that these definitions do not take uniformity into account: f_n is not required to be "uniformly describable" in dependence on n, nor do the straight-line programs Γ_n need to have anything in common. (But in all interesting situations, this will be the case.) We could incorporate a uniformity assumption into this model, for instance by requiring that f_n or Γ_n be polynomial time computable by a BSS-machine over k. In our opinion, this would just complicate the theory, but not make it more meaningful. At this point, we would also like to mention the paper [100] by Skyum and Valiant, which develops a nonuniform Boolean complexity theory along the same lines as it is done here.

We remark that our model is very simple, as only straight-line computations are considered. So we do not have branchings **if then else** according to the outcome of comparison tests. In fact, it is rather astonishing that completeness results can be obtained in such a restricted framework!

One more comment: the restriction to p-bounded degrees is a severe one; although X^{2^n} can be computed with only n multiplications, the corresponding sequence is not considered to be p-computable, as the degrees grow exponentially.

We define now an analogue of the class NP. A p-family $f = (f_n)$ is called p-*definable* iff there exists a p-computable family $g = (g_n)$ such that for all n

$$f_n(X_1, \ldots, X_{v(n)}) = \sum_{e \in \{0,1\}^{u(n)-v(n)}} g_n(X_1, \ldots, X_{v(n)}, e_{v(n)+1}, \ldots, e_{u(n)}) \ .$$

The set of p-definable families form the complexity class VNP. The class VP is obviously contained in VNP, and *Valiant's hypothesis* claims that this inclusion is strict. Note that as in the BSS-setting, this hypothesis a priori depends on the underlying field k. This possible dependence will be investigated in Sect. 4.1.

We shall now define a reduction notion called p-projection, which will serve to compare p-families. A polynomial f_n is said to be a *projection* of a polynomial $g_m \in k[X_1, \ldots, X_u]$, for short $f_n \leq g_m$, iff $f_n(X_1, \ldots, X_{v(n)}) = g_m(a_1, \ldots, a_u)$ for some $a_i \in k \cup \{X_1, \ldots, X_{v(n)}\}$. That is, f_n can be derived from g_m through substitution by indeterminates and constants. Let us call a function $t \colon \mathbb{N} \to \mathbb{N}$ *p-bounded from above and below* iff there exists some $c > 0$ such that $n^{1/c} - c \leq t(n) \leq n^c + c$ for all n. We call a p-family $f = (f_n)$ a *p-projection* of $g = (g_m)$, in symbols $f \leq_p g$, iff there exists a function $t \colon \mathbb{N} \to \mathbb{N}$ which is p-bounded from above and below such that

$$\exists n_0 \ \forall n \geq n_0 : f_n \leq g_{t(n)} \ .$$

(This definition of $\leq_p$ differs slightly from the one given in [107].) Finally, a p-family $g \in \mathrm{VNP}$ is called VNP-*complete* iff any $f \in \mathrm{VNP}$ is a p-projection of g.

The main result of the theory is the following algebraic analogue of Thm. 1.1.

Theorem 1.3 (Valiant) *The p-family* PER *is* VNP-*complete if* char$k \neq 2$.

1.4 Overview of Main Results

We outline the organization of the book and present some of our highlights in an informal way.

Chap. 2 is devoted to a detailed introduction into Valiant's model. In particular, we give a complete and detailed proof of the VNP-completeness of the permanent family (Thm. 1.3).

In Chap. 3 we prove that the families of generating functions corresponding to various NP (or #P) complete graph properties are VNP-complete. We just mention here one particularly nice result. Let F and G be graphs and assume that F is connected. We define an F-factor of G as a spanning subgraph of G, all of whose connected components are isomorphic to F. For instance, if F is the connected graph K_2 with two nodes, then an F-factor is just a perfect matching of G. We assume now that G is the complete graph with n nodes and that all its edges e have a weight X_e, where the X_e are independent indeterminates. We then define the weight of an F-factor as the product of the weights of all its edges. The sum of the weights of all F-factors of G is called the *F-factor polynomial* $\mathrm{Fact}_n(F)$. Note the following: if we would work with the complete bipartite graph G and take $F = K_2$, then we would get the permanent polynomial PER_n.

In Thm. 3.16 we will show that the sequence of F-factor polynomials is VNP-complete if F has at least two nodes and char$k \neq 2$. The point is to prove that the permanent family is a p-projection of the family of F-factor polynomials. This can be shown with a similar technique as in Kirkpatrick and Hell [65].

We have discussed three theories of NP-completeness in this introduction. Each of them hinges on a fundamental hypothesis of type P $\neq$ NP, which seems to be a very hard mathematical problem. Can we at least say something about the interrelations between the different P-NP-hypotheses? More specifically, what are the connections of Cook's hypothesis with the BSS-hypothesis and Valiant's hypothesis over the complex numbers? Our knowledge about this is depicted in Fig. 1.1. It has been shown independently by several people, including ourselves, that the nonuniform version of Cook's hypothesis P/poly $\neq$ NP/poly implies P $\neq$ NP over $\mathbb{C}$. (For a proof see Cucker et al. [31]. A definition of these nonuniform complexity classes can be found in Sect. 4.3.)

Whether the corresponding implication is also true over the reals (where machines can branch according to $\leq$-test) is a major open problem in the BSS-theory.

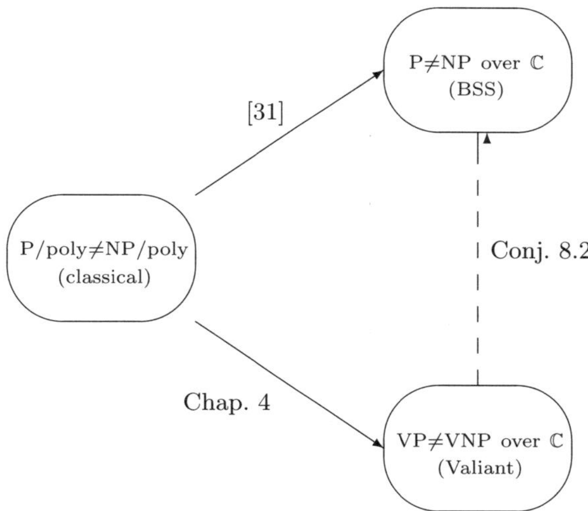

Figure 1.1: Interrelations between different P-NP-hypotheses.

In Chap. 4, we will prove that the nonuniform version of Cook's hypothesis also implies Valiant's hypothesis over $\mathbb{C}$, under the generalized Riemann hypothesis. In fact, a connection to parallel complexity shows up here: Valiant's hypothesis over $\mathbb{C}$ is also implied by the separation NC/poly $\neq$ P/poly. The main difficulty here is to eliminate the complex constants. For doing this, we have developed a general result about the frequency of primes p such that a system of integer polynomial equations solvable over $\mathbb{C}$ has a solution modulo p. We think this result (Thm. 4.4) is interesting in its own right.

We conjecture that Valiant's hypothesis implies the BSS hypothesis over the complex numbers. Loosely speaking, this means that if the permanent is intractable, then solving systems of polynomial equations is intractable as well. In Chap. 8 we will prove some weaker implications and clarify the problems that have to be overcome in order to settle this conjecture.

Chap. 5 is devoted to investigations in the spirit of structural complexity, similar as in Ladner [70] and Schöning [96] for the classical model. We call two p-definable families f, g p-equivalent iff we have $f \leq_p g$ and $g \leq_p f$. The equivalence classes form the poset of p-definable p-degrees. Assuming Valiant's hypothesis, we will prove that any countable poset can embedded in the poset of p-degrees. This proof will imply the existence of p-definable families of intermediate complexity: neither VNP-complete nor p-computable. Rather surprisingly, we found an explicit example for such a familiy over $\mathbb{F}_2$; the family of *cut enumerators* (Cut_n^2), which is defined as follows:

$$\mathrm{Cut}_n^2 := \sum_S \prod_{i \in A, j \in B} X_{ij} \ .$$

Here, the sum is over all cuts $S = \{A, B\}$ of the complete graph K_n on the set of nodes $\{1, 2, \ldots, n\}$. (The X_{ij} are independent indeterminates and we assume $X_{ji} = X_{ij}$. A cut of a graph is a partition of its set of nodes into two nonempty subsets.) We remark that in the classical and the BSS-model, one can also show the existence of problems of intermediate complexity, but no specific examples are known.

Chap. 6 and 7 are a continuation of work by Hartmann [47] and Barvinok [5] on the complexity of immanants.

From the viewpoint of computational complexity, determinant and permanent have, in spite of the similarity in their definition, very little in common. While the determinant can be evaluated with a polynomial number of arithmetic operations (Gaussian elimination), Valiant's completeness result indicates that this is not possible for the permanent. (The fastest known algorithm for the permanent due to Ryser [93] needs $O(m2^m)$ operations for an m by m matrix.)

Immanants are matrix functions which generalize the permanent and determinant. Let A be an m by m matrix and λ be a partition of m. The *immanant* of A corresponding to λ is defined as

$$\mathrm{im}_\lambda(A) = \sum_{\pi \in S_m} \chi_\lambda(\pi) \prod_{i=1}^m A_{i, \pi(i)} \ ,$$

where χ_λ is the irreducible character of the symmetric group S_m belonging to λ (cf. [13, 51]). For $\lambda = (1, \ldots, 1)$ this specializes to the determinant ($\chi_\lambda = \mathrm{sgn}$), and for $\lambda = (m)$ we obtain the permanent ($\chi_\lambda = 1$). For a deeper understanding of the amazingly different complexity behaviour of determinants and permanents, immanants are a natural object to study.

We have developed an efficient algorithm which evaluates the immanant im_λ with $O(m^2(m + s_\lambda d_\lambda))$ nonscalar operations (the total complexity is slightly bigger, see Thm. 7.3). Here, s_λ and d_λ denote the number of standard, and semistandard tableaus on the Young diagram of λ, respectively. So the invariant d_λ dominates the upper complexity bound. If this invariant is huge, the evaluation problem seems to become intractable. For instance, we have $d_\lambda = \binom{2m-1}{m}$ for permanents.

Our search for upper complexity bounds for immanants has lead us to the discovery of a fast algorithm to evaluate irreducible rational matrix representations of complex general linear groups with respect to a symmetry adapted basis (Gelfand-Tsetlin basis). This result (Thm. 6.6) is of considerable interest in its own right and may well have practical applications, given the importance of such representations in quantum mechanics. We confine ourselves with a simple application to the fast evaluation of associated Legendre functions.

In Chap. 7 we complement our algorithmic results with completeness proofs in Valiant's model for immanants corresponding to certain hook or rectangular diagrams. Even though these results clarify the picture and solve an open problem posed by Strassen [106], the complexity of immanants is still far from being completely understood.

In the next chapter, we shall now proceed with a thorough treatment of Valiant's model.

2

Valiant's Algebraic Model of NP-Completeness

We treat Valiant's algebraic theory of NP-completeness in detail and give a simplified VNP-completeness proof for the family of permanents. Then we discuss various closure properties of the complexity classes VP and VNP, in particular closedness with respect to factors. We also prove the important result on the efficient parallelization of straight-line programs due to Valiant, Skyum, Berkowitz, and Rackoff. As an application, we deduce the completeness of the determinant family in the class of qp-computable families.

2.1 The Complexity Classes VP and VNP

We first give a formal definition of the notion of complexity. In what follows, k denotes a fixed field and $X_1, X_2, \ldots$ are indeterminates over k.

Definition 2.1   (1) By a *straight-line program* Γ (expecting m inputs) we will understand a sequence $(\Gamma_1, \ldots, \Gamma_r)$ of instructions $\Gamma_\rho = (\omega_\rho; i_\rho, j_\rho)$ with operation symbols $\omega_\rho \in \{+, -, *\}$ and adresses i_ρ, j_ρ which are integers satisfying $-m < i_\rho, j_\rho < \rho$. We call r the *size* (or *length*) of Γ.

   (2) A straight-line program Γ defines a directed acyclic multigraph: its set of nodes is $\{\rho \in \mathbb{Z} \mid -m < \rho \leq r\}$ and it has the edges (i_ρ, ρ), (j_ρ, ρ). (An edge of multiplicity two occurs if $i_\rho = j_\rho$.) The number of edges of the longest directed path in this graph is called the *depth* of Γ.

   (3) For any sequence of input polynomials $a_1, \ldots, a_m$ a straight-line program Γ has a unique *result sequence* $(b_{-m+1}, \ldots, b_r)$ defined by $b_\rho = a_{m+\rho}$ for $\rho \leq 0$ and $b_\rho = b_{i_\rho} \omega_\rho b_{j_\rho}$ for $\rho > 0$. We say that Γ *computes* a set F of polynomials from the given inputs iff $F \subseteq \{b_{-m+1}, \ldots, b_r\}$.

Usually, the sequence of inputs is of the form $c_1, \ldots, c_s, X_1, \ldots, X_n$, with constants $c_i \in k$ and indeterminates X_j.

Definition 2.2 The *complexity* $L(F)$ of a set of polynomials $F \subseteq k[X_1, \ldots, X_n]$ is the minimal size of a straight-line program computing F from variables X_i and constants in k.

Let us fix some notation. We call a function $t\colon \mathbb{N} \to \mathbb{N}$ *p-bounded* iff there exists some $c > 0$ such that $t(n) \le n^c + c$ for all n. If we moreover have $n^{1/c} - c \le t(n) \le n^c + c$ for all n, then we will call the function t *p-bounded from above and below*.

Our main objects of study will be certain families (f_n) of polynomials.

Definition 2.3 A sequence $f = (f_n)$ of multivariate polynomials over k is called a *p-family* (over k) iff the number of variables as well as the degree of f_n are p-bounded functions of n.

Our assumption on the growth of degrees turns out to be crucial. Note that the sequence formed by the polynomials $f_n = X_1^{2^n}$ is not a p-family.

We provide now a formalization for the "feasible" families of polynomials.

Definition 2.4 A p-family $f = (f_n)$ is said to be *p-computable* iff the complexity $L(f_n)$ is a p-bounded function of n. The complexity class $\mathrm{VP} = \mathrm{VP}_k$ consists of all p-computable families over k.

It is important that this notion is insensitive to our particular choice of the complexity measure. For instance, we could have also allowed for divisions in our definition of complexity. However, over infinite fields k, a result due to Strassen [104] states that the complexity including divisions is a p-bounded function of $L(f)$ (division-free), the degree, and the number of variables of f. (Compare [21, Chap. 7].) Also, it is easy to see that the so-called nonscalar complexity measure, which only charges for the nonscalar multiplications, leads to the same complexity class VP.

The *determinant family* $\mathrm{DET} = (\mathrm{DET}_n)_n$ is an interesting example of a p-computable family. Here, DET_n denotes the determinant of an n by n matrix with independent indeterminate entries. Gaussian elimination shows that the complexity of DET_n is bounded by $O(n^3)$ when we allow for divisions. By the above remark, the determinant family is therefore contained in VP. (This is also true over finite fields, but requires additional justification.)

We introduce now another important class, that of p-definable polynomial families.

Definition 2.5 A p-family (f_n) is called *p-definable* iff there exists a p-computable family $g = (g_n)$, $g_n \in k[X_1, \dots, X_{u(n)}]$, such that for all n

$$ f_n(X_1, \dots, X_{v(n)}) = \sum_{e \in \{0,1\}^{u(n)-v(n)}} g_n(X_1, \dots, X_{v(n)}, e_{v(n)+1}, \dots, e_{u(n)}) \ . $$

The set of p-definable families form the complexity class $\mathrm{VNP} = \mathrm{VNP}_k$.

Observe that in general, f_n is obtained by a summation of exponentially many values of g_n, hence the complexity of f_n might be exponential in n. Note also that VP is clearly contained in VNP.

An interesting example of a p-definable family is provided by permanents. The *permanent* $\mathrm{per}(A)$ of an n by n matrix A is defined as

$$\mathrm{per}(A) = \sum_{\pi \in S_n} \prod_{i=1}^{n} A_{i,\pi(i)} \ .$$

This matrix function has a nice combinatorial interpretation: the permanent of the adjacency matrix of a bipartite graph equals the number of perfect matchings of this graph. The symbol PER_n shall denote the permanent of an n by n matrix of independent indeterminates. (Generally, we will adopt the following useful convention: we denote matrix functions with small letters, but the corresponding function evaluated at a matrix with independent indeterminate entries is written in capitals.)

Lemma 2.6 *The permanent family* $\mathrm{PER} = (\mathrm{PER}_n)$ *is p-definable.*

Proof. Let $X = (X_{ij})$ and $Y = (Y_{\ell m})$ be two $n \times n$ matrices of indeterminates. We define the polynomial g_n by

$$g_n(X,Y) := \underbrace{\left(\underbrace{\prod_{i,j,\ell,m} (1 - Y_{ij} Y_{\ell m})}_{=:\alpha_n(Y)} \right) \cdot \underbrace{\left(\prod_{i=1}^{n} \sum_{j=1}^{n} Y_{ij} \right)}_{=:\beta_n(Y)}}_{=:\gamma_n(Y)} \cdot \underbrace{\left(\prod_{i=1}^{n} \sum_{j=1}^{n} X_{ij} Y_{ij} \right)}_{=:\mu_n(X,Y)} ,$$

where the product in α_n is over all $1 \le i, j, \ell, m \le n$ such that $i = \ell$ iff $j \ne m$. As $L(g_n) = O(n^3)$, the family (g_n) is p-computable.

We claim that for all $e \in \{0,1\}^{n \times n}$, $\gamma_n(e)$ is nonzero iff e is a permutation matrix. Indeed, $\alpha_n(e) \ne 0$ iff every row and every column of e contain at most one 1. Moreover, if $\alpha_n(e) \ne 0$, then $\beta_n(e) \ne 0$ iff every row of e contains at least one 1. Altogether, $\gamma_n(e) = \alpha_n(e)\beta_n(e) \ne 0$ iff e is a permutation matrix. In this case we have $\mu_n(X, e) = \prod_{i=1}^{n} X_{i\pi(i)}$, where π denotes the permutation corresponding to e. Therefore, we get $\mathrm{PER}_n = \sum_{e \in \{0,1\}^{n \times n}} g_n(X, e)$, which proves that PER is p-definable. $\square$

A related family is $\mathrm{HC} = (\mathrm{HC}_n)$, the family of *Hamilton cycle polynomials*, defined by

$$\mathrm{HC}_n = \sum_{\pi} \prod_{i=1}^{n} X_{i,\pi(i)} \ ,$$

where the sum is over all cycles $\pi \in S_n$ of length n. Note that the value of HC_n at the adjacency matrix of a digraph equals the number of its Hamilton cycles. Similarly as in Lemma 2.6, we can show that HC is p-definable. This is not difficult, but somewhat cumbersome. In Sect. 2.3, we will learn a convenient criterion, which will immediately show the p-definability of most of the p-families we are interested in.

We will compare polynomials be means of a very simple reduction notion.

Definition 2.7 A polynomial f is called a *projection* of a polynomial g, written $f \le g$, iff

$$f(X_1, \ldots, X_n) = g(a_1, \ldots, a_m)$$

for some $a_i \in k \cup \{X_1, \ldots, X_n\}$. That is, f can be derived from g through substitution by indeterminates and constants.

It is clear that the projection $\le$ is transitive. We extend this to families of polynomials as follows.

Definition 2.8 We call a p-family $f = (f_n)$ a *p-projection* of $g = (g_m)$, in symbols $f \le_p g$, iff there exists a function $t \colon \mathbb{N} \to \mathbb{N}$ which is p-bounded from above and below such that

$$\exists n_0 \; \forall n \ge n_0 : f_n \le g_{t(n)} \; .$$

The reader should check that the p-projection is a transitive relation on the set of p-families. Our definition of $\le_p$ differs slightly from the one given by Valiant in [107]. On the one hand, we require the relation $f_n \le g_{t(n)}$ to hold for sufficiently large n only. In turn, in order to guarantee transitivity of $\le_p$, we have to make sure that $f(n) \to \infty$ as $n \to \infty$. For our purposes, it is convenient to achieve this by requiring that t is growing at least polynomially. Our modification of the definition will turn out to be useful in Chap. 5.

It is straightforward to check that the classes VP and VNP are closed under p-projection.

Definition 2.9 A p-definable family g is said to be VNP-*complete* iff $f \le_p g$ for all $f \in$ VNP.

We may view VNP-complete families as "hardest" members of the class VNP with respect to computational complexity. It is not obvious that VNP-complete problems do exist.

The following central theorem due to Valiant [107, 110] shows that PER is VNP-complete. This gives a partial explanation of the computational intractability of the permanents. (Note that permanents and determinants coincide in characteristic two.)

Theorem 2.10 (Valiant) *The family* HC *is* VNP-*complete. The same is true for the family* PER, *if* $\mathrm{char} k \ne 2$.

A simplified proof of this important result will be presented in the next section.

We have the following fundamental conjecture, which can be considered as an algebraic counterpart of the well-known hypothesis $\mathrm{P} \ne \mathrm{NP}$ due to Cook [25]. Its connection with hypotheses in classical complexity theory will be the topic of Chap. 4.

Valiant's hypothesis We have VP $\ne$ VNP.

Note that this hypothesis a priori depends on the underlying field k. This possible dependence will be investigated in Sect. 4.1.

The following remark is true since VP is closed under p-projections.

Remark 2.11 Valiant's hypothesis is true iff PER is not p-computable. The same holds true for any other VNP-complete family.

In Chap. 3 we will prove that the sequences of "generating functions" corresponding to various graph properties are VNP-complete. This will be achieved by proving that PER is a p-projection of each of these families. In particular, we will deduce the completeness of HC from the completeness of PER. (Of course, this only shows completeness for char $k \neq 2$.) We remark also that in Sect. 5.6, we will provide an existence proof for VNP-complete families, which does not rely on the intricate proof of Thm. 2.10.

The following lemma will be useful for completeness proofs later on. We call a p-family (f_n) *monotone* iff f_n is a projection of f_{n+1} for all n.

Lemma 2.12 *Let* $f = (f_n)$ *and* $g = (g_n)$ *be VNP-complete families and assume* f *to be monotone. Moreover, let* $I \subseteq \mathbb{N}$ *and define the corresponding "mixture"* $h = (h_n)$ *of* f *and* g *by setting* $h_n := f_n$ *if* $n \in I$ *and* $h_n := g_n$ *otherwise. Then* h *is VNP-complete as well.*

Proof. The p-family $(f_1, g_1, f_2, g_2, f_3, g_3, \ldots)$ is obviously p-definable and h is a p-projection of it, hence h is also p-definable.

Let $\varphi = (\varphi_n)$ be any p-definable family. Since f is complete, there is a function $n \mapsto t_1(n)$, which is p-bounded from above and below, such that $\varphi_n \leq f_{t_1(n)}$ for all sufficiently large n. We consider the p-definable family

$$\Phi_n := \varphi_n + Z^{1+D(t_1(n))} \ ,$$

where $D(n) := \max\{\deg g_i \mid 1 \leq i \leq n\}$ and Z is a new variable. As g is complete, there is a p-bounded $n \mapsto t_2(n)$ such that $\Phi_n \leq g_{t_2(n)}$. In particular,

$$D(t_1(n)) < \deg \Phi_n \leq \deg g_{t_2(n)} \ ,$$

which implies $t_2(n) > t_1(n)$. By substituting $Z \mapsto 0$ we see that $\varphi_n \leq g_{t_2(n)}$. On the other hand, we also have $\varphi_n \leq f_{t_2(n)}$, as f is monotone. From this it immediately follows that $\varphi_n \leq h_{t_2(n)}$. This shows that h is a complete family.
$\square$

We remark that the monotonicity assumption is necessary. Namely, if (f_n) is complete, then the families $(f_1, 0, f_2, 0, \ldots)$ and $(0, f_1, 0, f_2, \ldots)$ are both complete as well, but their mixture with respect to $I = \{2, 4, 6, 8, \ldots\}$ equals the zero sequence.

2.2   Completeness of the Permanent Family

Valiant originally developed his theory in a rather restricted framework, namely for formula size instead of complexity [107]. The importance of this theory became apparent later, when he showed that formula size may actually be replaced by complexity [110] (see Thm. 2.13 below). Valiant's original completeness proof for PER contained a gap. The first complete proof was published in von zur Gathen's survey article [41]. A different, matrix theoretical proof, was given in [21, Chap. 21]. Although the latter is very concise, we think the underlying ideas are best understood in terms of graph theory. We will therefore present here another demonstration of this important result, which is a simplification of the proof in von zur Gathen [41].

2.2.1   p-Definability and Formula Size

Formulas are inductively defined as follows: every variable X_i or constant in k is a formula; furthermore, if φ_1 and φ_2 are formulas, then so is $(\varphi_1 \circ \varphi_2)$, where $\circ \in \{+, *\}$. The *size* $E(\varphi)$ of a formula φ is the number of $+$ and $*$ used to build it. Every formula φ obviously represents a unique polynomial $\mathrm{val}(\varphi)$ in $k[X_1, \ldots, X_n]$. The *formula size* (or *expression size*) $E(f)$ of $f \in k[X_1, \ldots, X_n]$ is the smallest size of a formula φ with $\mathrm{val}(\varphi) = f$.

A formula for the polynomial f may be viewed as a special straight-line program, where intermediate results may be used only once: that is, the underlying acyclic digraph is a tree. In particular, we have $L(f) \leq E(f)$.

By replacing the complexity L by the formula size E in Def. 2.5, we get the complexity class VNP_e of *p-expressible families*. It is clear that $\mathrm{VNP}_e \subseteq \mathrm{VNP}$. Valiant [110] proved that these two classes actually coincide.

Theorem 2.13 *We have* $\mathrm{VNP}_e = \mathrm{VNP}$ *over any field.*

Our next goal is to provide the proof of this result. Recall the definition of a straight-line program Γ in Def. 2.1. Γ takes as inputs a sequence $c_1, \ldots, c_s, X_1, \ldots, X_n$ of constants $c_i \in k$ and indeterminates X_i and produces a result sequence (b_i). We think now of the constants c_i as being associated with Γ, and call Γ *homogeneous* iff all the intermediate results b_i are homogeneous polynomials. Recall also that Γ defines a directed acyclic multigraph: we will write $i \leq_\Gamma j$ iff there is a directed path in this graph from i to j. Note that the nonzero intermediate results b_i of a homogeneous straight-line program satisfy:

$$(2.1) \qquad\qquad i \leq_\Gamma j \implies \deg b_i \leq \deg b_j \ .$$

For a finite set F of polynomials, $H(F)$ shall denote the minimum length of a homogeneous straight-line program computing all the homogeneous parts of each $f \in F$. The number of variables of a polynomial f will be denoted by $v(f)$.

Lemma 2.14 *For polynomials f of degree d we have* $H(f) \leq (d+1)^2 L(f)$.

Proof. Denote the homogeneous part of degree δ of f by $f^{(\delta)}$. The proof follows easily from the following observation. If we have $b_\rho = b_s b_t$, then $b_j^{(\delta)} = \sum_{u=0}^{\delta} b_s^{(u)} b_t^{(\delta-u)}$. Hence we can compute the homogeneous parts of b_ρ up to degree d from the homogeneous parts of b_s and b_t up to degree d with $(d+1)^2$ arithmetic operations. If $b_\rho = b_s \pm b_t$, then we can perform this task with $d+1$ operations. $\square$

Proof. (of Thm. 2.13, taken from [21]) It is easy to see that the inclusion $\mathrm{VP} \subseteq \mathrm{VNP}_e$ implies $\mathrm{VNP} \subseteq \mathrm{VNP}_e$. Assume that $f = (f_n) \in \mathrm{VP}$. For proving $f \in \mathrm{VNP}_e$ we can w.l.o.g. assume by Lemma 2.14 that all f_n are homogeneous; furthermore we can restrict ourselves to homogeneous straight-line programs. So let $f_n \in k[X_1, \ldots, X_m]$, be homogeneous of degree d. We are going to construct recursively a family $g \in \mathrm{VP}_e$ such that $f_n(X) = \sum_e g_n(X, e)$ for all n. In a first step we describe and analyse the type of recursion. To this end let $P(C, d, m)$ denote the set of all homogeneous polynomials f in $k[X_1, \ldots, X_m]$ with $\deg f \le d$ and $H(f) \le C$. Furthermore define $A(C, d, m)$ as

$$\max_f \min\{E(h) \mid h \text{ homogeneous}, \; v(h) \le v(f) + 2C + 4, \; f(X) = \sum_e h(X, e)\} \;,$$

where the maximum is over all $f \in P(C, d, m)$. Obviously, $A(C, 1, m) \le 2m$ and $A(C, 2, m) \le 3(\binom{m}{2} + m) - 1 \le 3m^2$. We claim that the theorem is a consequence of the following recursion:

$$A(C, d, m) \le 3A(4C + \lfloor d/2 \rfloor, \lfloor d/2 \rfloor + 1, m + C) + \gamma C^2,$$

where γ is a suitable constant. To prove this claim, put $(C_0, d_0, m_0) := (C, d, m)$, and for $i \ge 1$ let

$$(C_i, d_i, m_i) := (4C_{i-1} + \lfloor d_{i-1}/2 \rfloor, \lfloor d_{i-1}/2 \rfloor + 1, m_{i-1} + C_{i-1}) \;.$$

Then an easy induction shows that for $2 \le 2^{\ell-1} < d_0 \le 2^\ell$ the following holds for all $i \le \ell$: $d_i \le 2^{\ell-i} + 1$; $d_\ell = 2$, $C_i \le 4C_{i-1} + 2^{\ell-i} \le 4^i C_0 + 2^{\ell-i} \frac{8^i - 1}{7}$, $m_i \le m_0 + \sum_{j=0}^{i-1} C_j$, and $A(C_0, d_0, m_0) \le 3^\ell A(C_\ell, 2, m_\ell) + \gamma \sum_{j=0}^{\ell-1} C_j^2$. Hence if C, d, and m are p-bounded functions in n, then so is the solution to the above recurrence. Hence all what remains to do is to deduce the above recurrence.

Assume $\Gamma = (\Gamma_1, \ldots, \Gamma_r)$ is an optimal homogeneous straight-line program computing $f_n \in P(C, d, m)$. Let (b_i) be the corresponding result sequence, and put $d_i := \deg b_i$. Write $\Gamma_j = (\omega_j; \alpha_j, \beta_j)$ and $\underline{r} := \{1, \ldots, r\}$.

Claim 1. If $J := \{j \in \underline{r} \mid \Gamma_j = (*; \alpha_j, \beta_j), \; d_j > d/2 \ge d_{\alpha_j}, d_{\beta_j}\}$, then there exists a homogeneous polynomial $S \in k[X_1, \ldots, X_m, E_0, E_j \mid j \in J]$ of degree $\lceil d/2 \rceil$ such that

(A) $H(S) \le H(f_n) + \lceil d/2 \rceil - 2.$

(B) $f_n(X) = \sum_{i \in J} b_i(X)S(X, e_0 + e_i)$, where e_i denotes the indicator function of $i \in J \cup \{0\}$.

Claim 1 will be proved by modifying Γ to a homogeneous straight-line program Γ' of length $H(f_n) + \lceil d/2 \rceil - 2$ that produces a result sequence (b'_i) with $d'_i := \deg b'_i$ such that for all $j \in \underline{r}$

(C)$_j$ $b_j = b'_j$, if $d_j \leq d/2$,

(D)$_j$ $b_j = \sum_{i \in J} b_i(X)b'_j(X, e_0 + e_i)$ and $d'_j = d_j - \lfloor d/2 \rfloor$, if $d_j > d/2$.

In particular, with $S := b'_r$ and the fact that $f_n = b_r$ we see that (A) and (B) hold.

Γ' is defined as follows. (In the sequel it will be convenient to deviate from the usual indexing of the instructions.) First of all we let Γ'_{0j} compute E_0^j, for $2 \leq j \leq \lceil d/2 \rceil - 1$. (This is possible in $\lceil d/2 \rceil - 2$ steps.) For $j \in \underline{r} \setminus J$ we put $\Gamma'_j := \Gamma_j$, and for $j \in J$ we let Γ'_j compute $E_0^{d_j - \lfloor d/2 \rfloor - 1}E_j$. For $\Gamma' = (\Gamma'_{02}, \ldots, \Gamma'_{0\lceil d/2 \rceil - 1}, \Gamma'_1, \ldots, \Gamma'_r)$, Claim (C)$_j$ holds obviously for all j. We prove (D)$_j$ for all j by induction on the partial ordering $<_\Gamma$ on the vertices of the directed acyclic multigraph corresponding to Γ. To prove the start we have to show that (D)$_j$ holds for all elements of J. So let $j \in J$. Then $b'_j(X, E) = E_0^{d_j - \lfloor d/2 \rfloor - 1}E_j$, $b'_j(X, e_0 + e_i) = \delta_{ij}$, for all $i \in J$, and b'_j is of degree $d_j - \lfloor d/2 \rfloor$. This settles the start. Now let $d_j > d/2$, but $j \notin J$. By the induction hypothesis, (C)$_\ell$ and (D)$_\ell$ are valid for all $\ell <_\Gamma j$.

Case 1. $\Gamma_j = (\pm; \alpha, \beta)$.
Since Γ is homogeneous we have $d_\alpha = d_\beta > d/2$, and as (D)$_\alpha$ and (D)$_\beta$ hold we obtain

$$
\begin{aligned}
b_j &= b_\alpha \pm b_\beta = \sum_{i \in J} b_i(X) \left(b'_\alpha(X, e_0 + e_i) \pm b'_\beta(X, e_0 + e_i) \right) \\
&= \sum_{i \in J} b_i(X) \underbrace{(b'_\alpha \pm b'_\beta)}_{=b'_j}(X, e_0 + e_i) \ .
\end{aligned}
$$

Finally, one easily checks that $d'_j = d_j - \lfloor d/2 \rfloor$.

Case 2. $\Gamma_j = (*; \alpha, \beta)$.
As $d_j > d/2$, but $j \notin J$ we have w.l.o.g. $d_\alpha > d/2 \geq d_\beta$. Hence, by (C)$_\beta$ and (D)$_\alpha$ and the fact that $b_\beta = b'_\beta$, we obtain

$$
\begin{aligned}
b_j &= b_\alpha * b_\beta = \sum_{i \in J} b_i(X)b'_\alpha(X, e_0 + e_i)b_\beta(X) \\
&= \sum_{i \in J} b_i(X) \underbrace{(b'_\alpha * b'_\beta)}_{=b'_j}(X, e_0 + e_i) \ .
\end{aligned}
$$

Furthermore, $d'_j = d'_\alpha + d_\beta = (d_\alpha - \lfloor d/2 \rfloor) + d_\beta = d_j - \lfloor d/2 \rfloor$ which proves (D)$_j$.

Claim 2. Let $J' := \{\alpha_j, \beta_j \mid j \in J\}$. Then the polynomial $T(X, V) := \sum_{i \in J'} b_i V_i V_0^{\lfloor d/2 \rfloor - d_i}$ is homogeneous of degree $\lfloor d/2 \rfloor + 1$ and $H(T) \leq H(f_n) + 3|J'| + \lfloor d/2 \rfloor - 3$. Furthermore, $b_i(X) = T(X, v_0 + v_i)$ for all $i \in J'$, where v_i is the indicator function of $i \in J' \cup \{0\}$.

In fact, a homogeneous straight-line program Γ'' with the properties proving Claim 2 is obtained by deleting from Γ every instruction Γ_i with $d_i > d/2$, adding further instructions that compute all powers V_0^i, for $2 \leq i < \lfloor d/2 \rfloor$, as well as all instructions that multiply each b_i, $i \in J'$, with $V_i V_0^{\lfloor d/2 \rfloor - d_i}$ to get the intermediate result z_i, and finally by adding further instructions to sum z_i for $i \in J'$.

Now consider the homogeneous polynomial $R(Z, V, V', E)$ defined by

$$ZV_0 V_0' E_0 \sum_{j \in J} V_{\alpha_j} V'_{\beta_j} E_j \prod_{i \in J' \setminus \{\alpha_j\}} (Z - V_i) \prod_{i \in J' \setminus \{\beta_j\}} (Z - V_i') \prod_{\rho \in J \setminus \{j\}} (Z - E_\rho) \ .$$

Note that R evaluated at a binary vector (z, v, v', e) of length $1 + 2(|J'| + 1) + (|J| + 1)$ gives 0 or 1. It is 1 iff $z = 1$ and there exists a (unique) index $j \in J$ such that $e = e_0 + e_j$, $v = v_0 + v_{\alpha_j}$, and $v' = v_0 + v_{\beta_j}$. Hence

(E) $$f_n(X) = \sum_{(z, v, v', e)} T(X, v) T(X, v') S(X, e) R(z, v, v', e) \ .$$

Recall that we are looking for a family $g \in \mathrm{VP}_e$ such that $f_n(X) = \sum_e g_n(X, e)$, for all n. We have already found a factor of g_n, namely R, which obviously has expression size at most $\gamma H(f_n)^2$, for some constant γ. The next step in the recursion processes $T(X, V)$ and $S(X, E)$. With (A), Claim 2, and $|J| + |J'| \leq H(f_n)$, we see that $4H(f_n) + \lfloor d/2 \rfloor$ is both an upper bound for $H(S)$ and $H(T)$. Thus using $J \neq \emptyset \neq J'$ Eq. (E) yields the stated recursion for $A(C, d, m)$. $\qquad\square$

2.2.2 Universality of the Permanent

Let $G = (V, E)$ be a digraph with two distinguished nodes, a source s and a sink t. We write $|G|$ for the number of nodes of G. By an s-t-path π we understand a simple directed path between s and t. Assume that a weight function $w \colon E \to k \cup \{X_1, X_2, \dots\}$ is given. The weight $w(\pi)$ of an s-t-path π is then defined as the product of the weights of the edges of π. We associate with G the value $\mathrm{SW}(G)$ defined as the sum of the weights of all s-t-paths of G (compare Sect. 3.3.6).

Lemma 2.15 *For any polynomial $f \in k[X_1, \dots, X_n]$ of formula size e there exists an acyclic, edge weighted digraph with $e + 2$ nodes such that $f = \mathrm{SW}(G)$. We can moreover assume that G has at most $e + 1$ edges of weight 1.*

Proof. Let φ be a formula of size e. By induction on e, we are going to construct an edge weighted digraph G such that $\mathrm{val}(\varphi) = \mathrm{SW}(G)$ and $|G| \leq e + 2$. The start $e = 0$ is clear.

Assume that $\varphi = (\varphi_1 + \varphi_2)$. We have $E(\varphi) = E(\varphi_1) + E(\varphi_2) + 1$. Let G_i be the digraph associated with φ_i by induction. We take the disjoint union of G_1 and G_2 and identify the sources and sinks of the G_i, respectively. Then the resulting digraph G clearly satisfies $SW(G) = SW(G_1) + SW(G_2)$. Moreover, $|G| = |G_1| + |G_2| - 2 \leq (E(\varphi_1) + 2) + (E(\varphi_2) + 2) - 2 = E(\varphi) + 1$.

Assume now that $\varphi = (\varphi_1 * \varphi_2)$. Again, $E(\varphi) = E(\varphi_1) + E(\varphi_2) + 1$. We take now the disjoint union of the digraphs G_i associated with φ_i and identify the sink of G_1 with the source of G_2. An s-t-path π of the resulting digraph G is given by a pair π_1, π_2 of s-t-paths of G_1, G_2, respectively, and we have $w(\pi) = w(\pi_1)w(\pi_2)$. This easily implies $SW(G) = SW(G_1)SW(G_2)$. Moreover, we have $|G| = |G_1| + |G_2| - 1 \leq E(\varphi) + 2$.

The statement about the number of edges with weight different from 1 is easily verified. $\qquad\square$

Assume G is an edge weighted digraph. After ordering the set of nodes of G, we can identify G with the matrix A, where A_{ij} is the weight of the edge pointing from the ith node to the jth node. (If there is no such edge, we set $A_{ij} = 0$.) We define the permanent $\mathrm{per}(G)$ of G as the permanent of the matrix A. This is well defined, as the permanent is clearly invariant under permutations of rows or columns. A permutation of the nodes of G has a unique decomposition into cycles and can thus be identified and visualized as a *cycle cover* of the digraph G: this is a set of directed cycles of G whose node sets form a partition of the node set of G. We define the weight $w(\sigma)$ of a cycle cover σ as the product of the weights of the edges occuring in σ. Using this notation, we see that the permanent of G is the sum of the weights of all cycle covers of G. Note also that $\mathrm{per}(G)$ is a projection of $\mathrm{PER}_{|G|}$ if the weights of G are either variables or indeterminates. We will elaborate more on this graph theoretical interpretation of matrix functions in Chap. 3.

Proposition 2.16 *For any polynomial $f \in k[X_1, \ldots, X_n]$ of formula size e there is a square matrix A of size $e + 1$ over $k \cup \{X_1, \ldots, X_n\}$ such that $f = \mathrm{per}(A)$. Moreover, we can assume that A has at most $e + 1$ entries different from 1.*

Proof. Let G be the digraph from Lemma 2.15. We identify s with t and introduce loops of weight 1 at all nodes different from $s = t$. Let us denote the resulting edge weighted digraph by H. We have $|H| = |G| - 1 \leq e + 1$. To any s-t-path π of G there corresponds a cycle cover σ of H consisting of loops and of the cycle obtained from π after identifying s with t. This correspondence is bijective since G is acyclic. Clearly, $w(\pi) = w(\sigma)$. We therefore obtain $f = SW(G) = \mathrm{per}(H)$. Any matrix A corresponding to H has the desired properties. $\qquad\square$

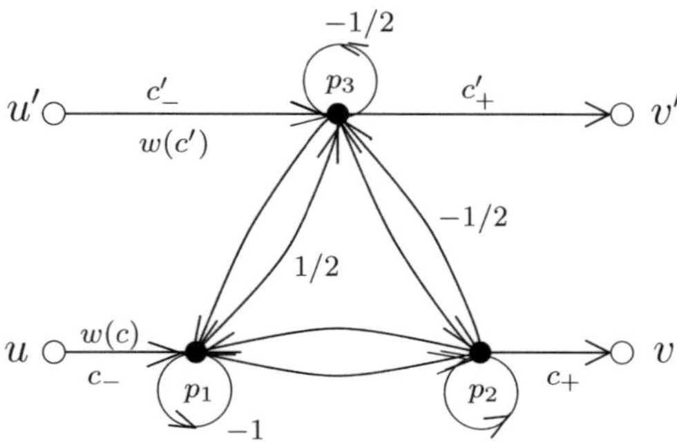

Figure 2.1: Iff-coupling of the edges $c = (u, v)$ and $c' = (u', v')$.

2.2.3 Completeness of the Permanent

Our goal is to provide a proof of Thm. 2.10 for the permanent family. By Lemma 2.6, we know already that PER is p-definable. The completeness follows immediately from the characterization of VNP in terms of formula size given in Thm. 2.13 and the next proposition. We continue using the graph theoretical interpretation of matrices introduced in the last subsection.

Proposition 2.17 *Assume the polynomial $g \in k[X_1, \ldots, X_n, Y_1, \ldots, Y_m]$ has formula size $E(g) < e$ and put*

$$f(X) = \sum_{e \in \{0,1\}^m} g(X, e) \ .$$

Then there exists a digraph F' of size $|F'| \leq 6e$ with edge weights in $k \cup \{X_1, \ldots, X_n\}$ such that $f = \mathrm{per}(F')$. In other words, f is a projection of PER_{6e}.

For the proof we will need two auxiliary constructions.

Suppose we have an edge weighted digraph G with distinguished edges $c = (u, v)$ and $c' = (u', v')$. We assume that u, v, u', v' are pairwise distinct. We insert between c and c' an auxiliary digraph K according to Fig. 2.1. (In this figure, edges have weight 1 unless otherwise indicated.) That is, we introduce two additional nodes p_1, p_2 on c and one additional node p_3 on c' and connect the nodes p_i according to the following weight matrix (also denoted by K)

$$K := \begin{pmatrix} -1 & 1 & \frac{1}{2} \\ 1 & 1 & -\frac{1}{2} \\ 1 & 1 & -\frac{1}{2} \end{pmatrix} \ .$$

Moreover, the edge $c_- := (u, p_1)$ is given the weight $w(c)$ formerly carried by c. Similarly, $c'_- := (u', p_3)$ has now the weight $w(c')$. The edges $c_+ := (p_2, v)$ and

$c'_+ := (p_3, v')$ are supposed to be of weight 1. The resulting digraph is denoted by $\overline{G}$. We will say that $\overline{G}$ results from G by *iff-coupling* of the edges c and c'. The next lemma clearly explains this naming.

Lemma 2.18 *The permanent of $\overline{G}$ equals the sum of the weights of all cycle covers of G, which contain either both of the edges c and c', or none.*

Proof. The matrix K has the following nice properties, which play a key role in our construction. If $K[R|C]$ denotes K with rows in R and columns in C removed, then we have:

$$\mathrm{per}(K[2|1]) = \mathrm{per}(K[2|3]) = \mathrm{per}(K[3|1]) = \mathrm{per}(K[3|3]) = 0$$

and

$$\mathrm{per}(K) = \mathrm{per}(K[2,3|1,3]) = 1 \ .$$

For $C \subseteq \mathcal{E} := \{c_-, c_+, c'_-, c'_+\}$ let $S(C)$ be the set of all cycle covers $\overline{\sigma}$ of $\overline{G}$ which, among the edges in $\mathcal{E}$, contain exactly the edges in C. There are at most 6 possibilities for C with nonempty $S(C)$:

$$C_1 = \{c_-, c_+\}, C_2 = \{c'_-, c_+\}, C_3 = \{c_-, c'_+\}, C_4 = \{c'_-, c'_+\}, C_5 = \emptyset, C_6 = \mathcal{E} \ .$$

Let W_i denote the sum of the weights of the cycle covers in $S(C_i)$. Then we have $\mathrm{per}(\overline{G}) = W_1 + \ldots + W_6$.

We are going to show that $W_1 = 0$. Each cycle cover σ of G containing c, but not c', can be "extended" to exactly two cycle covers $\overline{\sigma}_1, \overline{\sigma}_2$ in $S(C_1)$, and all cycle covers in $S(C_1)$ are obtained in this way. We have

$$w(\overline{\sigma}_1) + w(\overline{\sigma}_2) = w(\sigma)[-\frac{1}{2} + \frac{1}{2}] = 0 \ .$$

The number in square brackets has the following interpretation: it is the sum of the weights of all cycle covers of K, which do not have an incoming edge to p_1 and an outcoming edge from p_2. In other words, this number equals the permanent $\mathrm{per} K[2|1] = 0$. We conclude that $W_1 = 0$.

In a similar way, using the above mentioned properties of K, we can show that $W_2 = \ldots = W_4 = 0$. W_5 turns out to be the sum of the weights of all cycle covers of G which neither contain c nor c', and W_6 is the sum of the weights of all cycle covers of G which contain both c and c'. This proves the lemma. $\qquad\square$

In our proof of Prop. 2.17 we need further auxiliary digraphs: the *rosettes* $R(\mu)$ defined for a positive integer μ (cf. Fig. 2.2). They are constructed as follows. We start with a directed cycle of length μ, consisting of the nodes $u_1, \ldots, u_\mu$ and the *connector edges* $c_i = (u_i, u_{i+1})$. (Here and in the following, the index arithmetic is supposed to be modulo μ.) We introduce additional nodes $v_1, \ldots, v_\mu$ together with the edges $(u_i, v_i), (v_i, u_{i+1})$. Finally, we add

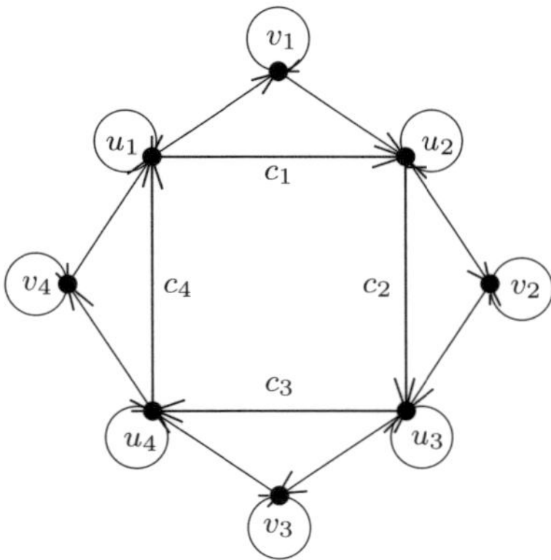

Figure 2.2: The rosette $R(4)$.

loops (u_i, u_i) and (v_i, v_i) at all the nodes. The resulting digraph will be called
the rosette $R(\mu)$. All edges of $R(\mu)$ are supposed to carry the weight 1. We
denote by $C = \{c_1, \ldots, c_\mu\}$ the set of connector edges of $R(\mu)$.

The rosette $R(\mu)$ has the following easily verified properties.

- For each nonempty subset $S \subseteq C$ there is exactly one cycle cover of $R(\mu)$,
  which, among the connector edges, contains exactly the edges in S.

- There are exactly two cycle covers of $R(\mu)$, which contain no connector edge. One of them consists of loops only: it will be called the *loop
  covering* of $R(\mu)$.

Proof. (of Prop. 2.17) By Prop. 2.16 there is a digraph G with $|G| \leq e$ such
that $g = \mathrm{per}(G)$. The weights of the edges of G are variables X_i, Y_j or constants in k. Let $d_{i,1}, \ldots, d_{i,\mu_i}$ be the edges of G carrying the weight Y_i. By
Prop. 2.16, we have $\mu_1 + \ldots + \mu_m \leq e$. We may assume that $\mu_i > 0$ for all i.

We form the disjoint union F of the digraph G and the rosettes $R(\mu_i)$ for
$1 \leq i \leq m$. The connector edges of $R(\mu_i)$ are denoted by $c_{i,j}$ for $1 \leq j \leq \mu_i$.
The digraph F inherits its edge weights w_F from G and the rosettes, except
that we give all the edges $d_{i,j}$ the new weight 1.

Let Φ denote the set of cycle covers of F which, for all i, j, contain the
edge $c_{i,j}$ iff they contain the edge $d_{i,j}$. Such cycle covers are said to respect
iff-couplings of the edges $c_{i,j}$ with $d_{i,j}$. We claim that it suffices to prove that

$$(2.2) \qquad f(X) = \sum_{\varphi \in \Phi} w_F(\varphi) \ .$$

In fact, we can realize all the iff-couplings between $c_{i,j}$ and $d_{i,j}$ by means of
the auxiliary construction of Fig. 2.1. Let $\overline{F}$ be the edge weighted digraph
thus resulting from F. By Lemma 2.18 we have $\mathrm{per}(\overline{F}) = \sum_{\varphi \in \Phi} w_F(\varphi)$. From

(2.2) we can then conclude that $\mathrm{per}(\overline{F}) = f$. Moreover, we have

$$|F'| \le |G| + 2\sum_{i=1}^{m} \mu_i + 3\sum_{i=1}^{m} \mu_i \le 6e \ .$$

This will prove the proposition.

It remains to verify (2.2). Let Γ be the set of cycle covers of G. A cycle cover $\varphi \in \Phi$ restricted to G and $R(\mu_i)$ yields cycle covers $\gamma = \varphi_{|G}$ of G and $\varphi_{|R(\mu_i)}$ of $R(\mu_i)$, respectively. We associate with φ the following subset $J(\varphi)$ of $\underline{m} := \{1, \dots, m\}$:

$$J(\varphi) := \{i \in \underline{m} \mid \varphi_{|R(\mu_i)} \text{ is the loop covering of } R(\mu_i)\} \ .$$

Moreover, we associate with a cycle cover $\gamma \in \Gamma$ the following subset $I(\gamma)$ of $\underline{m}$, defined by

$$I(\gamma) := \{i \in \underline{m} \mid \exists j \ \gamma \text{ contains edge } d_{i,j}\} \ .$$

Note that if $\varphi \in \Phi$ and $\gamma = \varphi_{|G}$, then $I(\gamma)$ and $J(\varphi)$ are disjoint. Indeed, if γ contains $d_{i,j}$, then φ contains $c_{i,j}$, since φ is supposed to respect the iff-coupling of these edges. On the other hand, this implies that the restriction of φ to $R(\mu_i)$ cannot be the loop covering of $R(\mu_i)$.

We thus get a mapping $\varphi \mapsto (\varphi_{|G}, J(\varphi))$ from Φ to the set of pairs (γ, J), where $\gamma \in \Gamma$ and $J \subseteq \underline{m}$ is disjoint to $I(\gamma)$. By the properties of rosettes and iff-couplings, this is a bijection. Indeed, if $i \in I(\gamma)$, then an extension $\varphi \in \Phi$ of γ must contain some connector edge $c_{i,j}$, hence the extension to the rosette $R(\mu_i)$ is uniquely determined. If $i \notin I(\gamma)$, then there are a priori two extensions to $R(\mu_i)$. The additional requirement $J = J(\varphi)$ makes the choice unique.

To a subset $J \subseteq \underline{m}$ we assign a modified weight function w_G^J of G, which results from w_G by the substitution $Y_i \mapsto 0$ if $i \in J$, and $Y_i \mapsto 1$ if $i \notin J$. That is, the edges $d_{i,j}$ get the weight 0 if $i \in J$, and the weight 1 if $i \notin J$.

Assume $\varphi \in \Phi$ and $\gamma = \varphi_{|G}$. If $I(\gamma)$ is disjoint from J, then φ contains none of the edges $d_{i,j}$ with $i \in J$. Therefore, $w_G^J(\gamma) = w_F(\varphi)$ in this case. On the other hand, if there is some $i \in I(\gamma) \cap J$, then φ contains some $d_{i,j}$. As $w_G^J(d_{i,j}) = 0$, we have $w_G^J(\gamma) = 0$ in that case.

A subset $J \subseteq \underline{m}$ can be identified with the vector $e \in \{0,1\}^m$ characterized by $e_i = 0$ iff $i \in J$. From $g(X,Y) = \mathrm{per}(G) = \sum_{\gamma \in \Gamma} w_G(\gamma)$ we conclude that

$$g(X, e) = \sum_{\gamma \in \Gamma} w_G^J(\gamma) \ .$$

This implies

$$f(X) = \sum_{e \in \{0,1\}^m} g(X, e) = \sum_{J \subseteq \underline{m}} \sum_{\gamma \in \Gamma} w_G^J(\gamma) \ .$$

By the above observation, we have for $\gamma = \varphi_{|G}, \varphi \in \Phi$ that

$$\sum_{J \subseteq \underline{m}} w_G^J(\gamma) = \sum_{J \cap I(\gamma) = \emptyset} w_G^J(\gamma) = \sum_{J \cap I(\gamma) = \emptyset} w_F(\varphi) \ .$$

This implies

$$f(X) = \sum_{\gamma \in \Gamma} \sum_{J \subseteq \underline{m}} w_G^J(\gamma) = \sum_{\gamma \in \Gamma} \sum_{J \cap I(\gamma) = \emptyset} w_F(\varphi) = \sum_{\varphi \in \Phi} w_F(\varphi) \ ,$$

which proves equation (2.2). $\qquad\qquad\qquad\qquad\qquad\qquad\qquad\qquad\qquad$ $\square$

2.3 Closure Properties

The complexity class VNP is closed under various natural operations. We have the following result due to Valiant [110].

Theorem 2.19 *Let* $(f_n), (g_n)$ *be p-definable, say* $f_n \in k[X_1, \ldots, X_{v(n)}]$. *Then:*

(1) *(Sum and product)* $(f_n + g_n)$ *and* $(f_n \cdot g_n)$ *are p-definable.*

(2) *(Substitution)* $(f_n(g_1, \ldots, g_{v(n)}))$ *is p-definable.*

(3) *(Coefficient) If* $h_n \in k[X_{u(n)+1}, \ldots, X_{v(n)}]$ *is the coefficient of some power product* $X_1^{i_1} \cdots X_{u(n)}^{i_{u(n)}}$ *in* f_n *for some* $u(n) \leq v(n)$, *then the family* (h_n) *is p-definable.*

It is clear that the class VP is also closed under the formation of sums, products, as well as substitutions. However, the class VP is not closed under the operation of taking coefficients, if VP $\neq$ VNP. This can be seen from the following example. Consider

$$f_n := \prod_{i=1}^{n} \left(\sum_{j=1}^{n} X_{ij} Y_j \right) \ .$$

The family (f_n) is clearly p-computable, as $L(f_n) = O(n^2)$. On the other hand, the coefficient of the product $Y_1 \cdots Y_n$ in f_n equals the permanent PER_n.

We remark that Valiant has also proved in [110] that VNP is closed under p-bounded applications of differentiation and integration. Again, the class VP fails to be closed under these operations, as the following formula shows:

$$\frac{\partial}{\partial Y_1} \cdots \frac{\partial}{\partial Y_n} f_n = \left(\frac{3}{2} \right)^n \int_{-1}^{1} \cdots \int_{-1}^{1} Y_1 \cdots Y_n f_n \, dY_1 \cdots dY_n = \mathrm{PER}_n \ .$$

We discuss now a useful criterion for p-definability due to Valiant [107]. Recall the counting complexity class #P from the introduction. This class #P is contained in its nonuniform version #P/poly, for whose definition we refer to Sect. 4.3.

Proposition 2.20 (Valiant's criterion) *Suppose $\phi\colon \{0,1\}^* \to \mathbb{N}$ is a function in the class $\#P/\text{poly}$. Then the family (f_n) of polynomials defined by*

$$f_n = \sum_{e \in \{0,1\}^n} \phi(e) X_1^{e_1} \cdots X_n^{e_n}$$

is p-definable.

Proof. Assume first that ϕ is in $\#P$. By definition, there is a balanced, polynomial time decidable string relation R with $\phi(e) = \#\{y \mid (x,y) \in R\}$. As in the NP-completeness proof for the satisfiability problem for Boolean formulas one shows that for each $n \in \mathbb{N}$ there is a 3-conjunctive normal form Φ_n in the variables $E_1, \ldots, E_n, Y_1, \ldots, Y_{m(n)}$, having $n^{O(1)}$ clauses, such that $m(n) = n^{O(1)}$ and

$$\phi(e) = \#\{y \in \{0,1\}^{m(n)} \mid \Phi_n(e,y) \text{ true}\}$$

for all $e \in \{0,1\}^n$ (compare [87, §8.2]). It is easy to see that for each 3-clause K there is an integer polynomial g_K in at most three of the variables E_i, Y_j having degree at most 3 and such that

$$\forall x \in \{0,1\}^3 : g_K(x) = \begin{cases} 1 & \text{if } K(x) \text{ true,} \\ 0 & \text{otherwise.} \end{cases}$$

(For instance, for $K = u \vee v \vee w$ take $g_K := uvw + uv(1-w) + u(1-v)w + (1-u)vw + u(1-v)(1-w) + (1-u)v(1-w) + (1-u)(1-v)w$.)

Let p_n be the product of the g_K over all clauses K of Φ_n. Clearly, (p_n) is p-computable. Moreover, we have for all $e \in \{0,1\}^n$ that

$$\phi(e) := \sum_{y \in \{0,1\}^{m(n)}} p_n(e,y) \ .$$

Let now X_i be additional variables and define the polynomial

$$h_n(X,E,Y) := p_n(E,Y) \prod_{i=1}^{n} (E_i X_i + 1 - E_i) \ .$$

The family (h_n) is p-computable and we have

$$f_n = \sum_{e \in \{0,1\}^n} \phi(e) X_1^{e_1} \cdots X_n^{e_n} = \sum_{e \in \{0,1\}^n} \sum_{y \in \{0,1\}^{m(n)}} h_n(X,e,y) \ ,$$

which shows that (f_n) is p-definable.

Essentially the same argument works for functions ϕ in $\#P/\text{poly}$. $\qquad\square$

The remainder of this section is dedicated to the proof that the class VP is closed with respect to factors.

$M(d)$ will denote an upper bound on the complexity for the symbolic multiplication of two polynomials of degree d over k. By symbolic multiplication, we understand the computation of the coefficients of the product polynomial from the coefficients of the given factors. It is well known that $M(d) = O(d \log d)$ if the field k "supports" fast Fourier transforms, for instance, if k is the field of complex numbers (cf. [21, Sect. 2.1]). As our main concern is in polynomial complexity bounds, the reader less interested in details may think of the trivial bound $M(d) = O(d^2)$.

Kaltofen [57, 58] has designed a probabilistic reduction of the problem to factor multivariate polynomials given by straight-line programs to the bivariate factorization problem. The proof of the theorem below is similar as in this paper, but considerably simpler, since we only claim the existence of short straight-line programs and do not care about their construction. We adopt the particularly elegant viewpoint of Kaltofen and Trager [59], which interprets Hensel lifting as a homotopy continuation method with respect to a discrete valuation. (Homotopy continuation is a standard method in numerical analysis for solving nonlinear systems of equations, see for instance [2].)

Throughout the rest of this section, we assume that k is a field of characteristic zero.

Theorem 2.21 *Let $f \in k[X_1, \ldots, X_n]$ be a multivariate polynomial of degree d. There is a division-free straight-line program which computes all the irreducible factors of f with $(d+1)^2$ evaluations of f and $O(dM(d)^2 \log d + dn)$ arithmetic operations. Thus any factor g of f has a complexity*

$$L(g) \le (d+1)^2 L(f) + O(dM(d)^2 \log d + dn) \ .$$

In particular, the complexity of g is polynomially bounded in the complexity of f, the degree d, and the number of variables n.

The main conclusion of this theorem can be succinctly stated in terms of the complexity class VP^f of p-computable families relative to f, that we will be introduced in Sect. 5.6.

Corollary 2.22 *Let $f = (f_n)$ and $g = (g_n)$ be p-families and assume that g_n is a factor of f_n for all n. Then $g \in \mathrm{VP}^f$. In particular, the class of p-computable families is closed under taking factors: $f \in \mathrm{VP}$ implies $g \in \mathrm{VP}$.*

Proof. (of Thm. 2.21) It is convenient to write $X := X_1$ and $Y_i := X_{i+1}$ for $i > 1$. Assume that $f = \prod_{i=1}^{r} g_i^{e_i}$ is the factorization of f into irreducible polynomials g_i. By a linear transformation $X \mapsto X$, $Y_i \mapsto Y_i + \alpha_i X$ we can achieve that $\deg_X f = \deg f = d$. We may therefore assume that f and all g_i are monic with respect to X. In what follows, we will consider f and g_i

as univariate polynomials in X over the coefficient ring $R := k[Y_1, \ldots, Y_{n-1}]$. Put $d_i := \deg g_i = \deg_X g_i$.

The g_i are pairwise relatively prime. Hence there is some $y \in k^{n-1}$ such that the polynomials $g_{i,0} := g_i(X, y)$ are relatively prime as well. Note that $\deg g_{i,0} = d_i$. We have the factorization $f_0 := f(X, y) = \prod_i g_{i,0}^{e_i}$.

As in the papers [43, 57], the basic idea is to use Hensel lifting in order to successively construct the factorization of f from those of f_0. As mentioned in [59], we may interpret this process as a homotopy continuation method with respect to a discrete valuation. Let T be a new indeterminate and consider the "segment"

$$a(X, T) := (X, TY + (1 - T)y)$$

connecting (X, y) with the identity (X, Y). We set

$$F := f(a(X, T)) = \sum_{j=0}^{d} f_j T^j \ , \ \ G_i := g_i(a(X, T)) = \sum_{j=0}^{d_i} g_{i,j} T^j \ ,$$

with $f_i, g_{i,j} \in R[X]$. This notation is consistent with our earlier introduction of f_0 and $g_{i,0}$. Note that $\deg g_{i,j} < d_i$ for $j > 0$, and $g_i = \sum_{j=0}^{d_i} g_{i,j}$ (substitute T by 1).

We will now derive a formula, which will allow to compute the $g_{i,s+1}$ from the $g_{i,0}, \ldots, g_{i,s}$ $(i \leq r)$ and $f_0, \ldots, f_s$. We set

$$\sum_{j \geq 0} \pi_{s,j} T^j := \prod_{i=1}^{r} \left(\sum_{j=0}^{s} g_{i,j} T^j \right)^{e_i} \ .$$

From $F = \prod_{i=1}^{r} G_i^{e_i}$ we obtain modulo T^{s+2} that

$$\begin{aligned}
F &\equiv \prod_{i=1}^{r} \left(\sum_{j=0}^{s} g_{i,j} T^j + g_{i,s+1} T^{s+1} \right)^{e_i} \\
&\equiv \prod_{i=1}^{r} \left(\left(\sum_{j=0}^{s} g_{i,j} T^j \right)^{e_i} + e_i g_{i,0}^{e_i-1} g_{i,s+1} T^{s+1} \right) \\
&\equiv \prod_{i=1}^{r} \left(\sum_{j=0}^{s} g_{i,j} T^j \right)^{e_i} + \sum_{i=1}^{r} e_i g_{i,0}^{e_i-1} g_{i,s+1} \prod_{\ell \neq i} g_{\ell,0}^{e_\ell} \ T^{s+1} \ .
\end{aligned}$$

This implies

$$(2.3) \qquad \frac{v_{s+1}}{\prod_i g_{i,0}} = \sum_i \frac{e_i g_{i,s+1}}{g_{i,0}} \ ,$$

with the polynomial

$$v_{s+1} := \frac{f_{s+1} - \pi_{s,s+1}}{\prod_i g_{i,0}^{e_i-1}} \ .$$

Since the $g_{i,0}$ are pairwise coprime and $\deg g_{i,s+1} < d_i = \deg g_{i,0}$, (2.3) is the uniquely determined partial fraction decomposition of the rational function on the left-hand side. Consider the partial fraction decomposition

$$\frac{1}{\prod_i g_{i,0}} = \sum_{i=1}^{r} \frac{h_i}{g_{i,0}} \ ,$$

where $h_i \in k[X]$ with $\deg h_i < d_i$. We obtain from this

$$\frac{v_{s+1}}{\prod_i g_{i,0}} = A + \sum_i \frac{h_i v_{s+1} \bmod g_{i,0}}{g_{i,0}}$$

for some polynomial A. The uniqueness of the partial fraction decomposition implies $A = 0$ and that (here we use $\operatorname{char} k = 0$)

$$g_{i,s+1} = e_i^{-1} h_i v_{s+1} \bmod g_{i,0} \ .$$

We describe now the straight-line program for computing the g_i from the Y_i, X and constants in k. Note that the coefficients of $g_{i,0}, h_i$ as well as the components of y are constants in k. The cost of actually producing these values is not taken into account in our model.

(1) We choose distinct points $x_0, \ldots, x_d, t_0, \ldots, t_d$ in k and compute $a(x_\rho, t_\sigma) = (x_\rho, t_\sigma Y + (1 - t_\sigma)y)$ for $0 \le \rho, \sigma \le d$ with $O(dn)$ operations. From this we get $f(a(x_\rho, t_\sigma))$ by $(d+1)^2$ evaluations of f. The coefficients of a univariate polynomial of degree d can be computed from its values at $d+1$ distinct points using $O(M(d) \log d)$ operations, cf. [21, Cor. 3.22]. Therefore, we may get the coefficients of $f_0, \ldots, f_d$ from the $f(a(x_\rho, t_\sigma))$ with $O(dM(d) \log d)$ arithmetic operations. (Note that we compute a set of elements of R.)

(2) Iterate the following procedure for $s = 0, \ldots, d - 1$:

Suppose we have already computed the coefficients of $g_{i,0}, \ldots, g_{i,s}$ for all $1 \le i \le r$. The algorithm in [21, Cor. 2.15] for multiplying several polynomials symbolically shows that we can compute the coefficients of $\pi_{s,s+1}$ using $O(M(d) \cdot M(d) \log d)$ operations. (For this, we spend $O(M(d) \log d)$ arithmetic operations on polynomials in $R[X]$ of degree at most d, each of which can be realized by $O(M(d))$ arithmetic operations in R.) We get the coefficients of $u_{s+1} := f_{s+1} - \pi_{s,s+1}$ with further $O(d)$ subtractions. As the co-efficients of $p := \prod_i g_{i,0}^{e_i - 1}$ are constants in k, we can compute the coefficients of $v_{s+1} = u_{s+1}/p$ with $O(M(d))$ arithmetic operations using the algorithm in [21, Cor. 2.26] (no divisions are necessary, since p_i is monic). Using the same algorithm r times, we may obtain the coefficients of $v_{s+1} \bmod g_{i,0}$ for all i with $O(rM(d))$ operations. (One can do even better: $O(M(d) \log d)$ operations suffice for this, see [21, Thm. 3.19].) From this we get the coefficients of $g_{i,s+1} = e_i^{-1} h_i v_{s+1} \bmod g_{i,0}$ for all i using $O(\sum_i M(d_i)) \le O(M(d))$ operations.

(3) We have now computed the coefficients of $g_{i,j}$ for $i \leq r, j < d$. From this we get all the $g_i = \sum_{j=0}^{d_i} g_{i,j}$ with $O(\sum_i d_i) \leq O(d)$ operations.

Altogether, we have computed all the g_i using $(d+1)^2$ evaluations of f and a total of $O(dM(d)^2 \log d + dn)$ arithmetic operations, without divisions. $\square$

Problem 2.1 Is the class VP also closed under taking factors over fields of positive characteristic?

Conjecture 2.1 The class VNP is closed under taking factors.

2.4   Parallel Complexity

We discuss here an important result on the efficient parallelization of straight-line programs due to Valiant, Skyum, Berkowitz, and Rackoff [113]. This result will be needed later in Chap. 4.

Theorem 2.23 (Valiant et al.) *Let f be an n-variate polynomial of degree d. Then f can be computed by a straight-line program of size $O(d^6 L(f)^3)$ and depth $O(\log(dL(f)) \log d + \log n)$.*

The essence of this theorem can be succinctly stated in terms of the following algebraic analogues of the complexity classes NC^i, known as Nick's classes (cf. Cook [26] and Pippenger [90]).

Definition 2.24 For a positive integer i let VNC^i be the class of all p-families (f_n) over k such that there is a sequence (Γ_n) of straight-line programs having the following properties: Γ_n computes f_n, the size of Γ_n is a p-bounded function of n, and the depth of Γ_n is $O(\log^i n)$.

Obviously, $\mathrm{VNC}^1 \subseteq \mathrm{VNC}^2 \subseteq \ldots \subseteq \mathrm{VP}$. Surprisingly, this hierarchy collapses, as is seen by Thm. 2.23.

Corollary 2.25 *We have $\mathrm{VP} = \mathrm{VNC}^2$ over any field.*

By contrast, it is conjectured that the classical complexity classes P and NC are different. We also remark that in the BSS-model, the separation of the corresponding classes can be obtained rather easily (cf. [28]).

Proof. (of Thm. 2.23, taken from [21]) We may assume that $f(0) = 0$. Think of an optimal homogeneous straight-line program Γ computing the homogeneous parts of f from constants in k and indeterminates. A moment's thought shows that in the course of the computation, a constant λ can only be involved in a multiplication. It will be convenient to interpret this as the result of a scalar multiplication instruction. We will therefore allow our straight-line program Γ to have additional instructions of type $\Gamma_j = (\lambda; i)$, causing the effect

$b_j = \lambda b_i$ in the result sequence. In turn, we can assume that the indeterminates X_i are the only inputs.

Assume now that $\Gamma = (\Gamma_1, \ldots, \Gamma_r)$ is such a modified optimal homogeneous straight-line program computing the homogeneous parts of f. This straight-line program defines a directed acyclic multigraph: we write $i \leq_\Gamma j$ iff there is a directed path in this graph from i to j. Let $b = (b_j)_{j \in N}$ with $N := \{j \mid -n < j \leq r\}$ be the result sequence of Γ. We partition the index set $\underline{r}$ as $\underline{r} = A \sqcup M \sqcup S$, where $\Gamma_j = (\pm; j', j'')$ for $j \in A$, $\Gamma_j = (*; j', j'')$ for $j \in M$, and $\Gamma_j = (\lambda_j; j')$ for $j \in S$. By Lemma 2.14 we can assume that $r \leq (d+1)^2 L(f)$. Furthermore, we may assume the following (cf. (2.1)):

(A) Each b_j is nonzero and homogeneous of degree d_j, and $i \leq_\Gamma j$ implies $d_i \leq d_j \leq d$ for all i, j.

(B) $d_{j'} \geq d_{j''}$ for all $j \in A \sqcup M$.

Let $B = B(\Gamma, b) = (b_{ij})$ be an upper triangular matrix in $k[X_1, \ldots, X_n]^{N \times N}$ with ones on the main diagonal, whose columns B_j are defined recursively in the following way (e_j denotes the indicator function of $j \in N$):

$$
\begin{aligned}
j \in N, j \leq 0 &\;\Rightarrow\; B_j := e_j \\
\Gamma_j = (\lambda_j; j') &\;\Rightarrow\; B_j := e_j + \lambda_j \cdot B_{j'} \\
\Gamma_j = (+; j', j'') &\;\Rightarrow\; B_j := e_j + B_{j'} + B_{j''} \\
\Gamma_j = (*; j', j'') &\;\Rightarrow\; B_j := e_j + b_{j''} \cdot B_{j'} \;.
\end{aligned}
$$

Before proceeding with the proof we illustrate the construction of the matrix B by an example. The following describes a homogeneous straight-line program $\Gamma = (\Gamma_1, \ldots, \Gamma_5)$ and its result sequence $(b_i)_{-2 \leq i \leq 5} = (X_1, X_2, X_3, b_0, b_1, \ldots, b_5)$:

$$
\begin{aligned}
\Gamma_1 &= (\lambda; -2) & b_1 &= \lambda X_1 \\
\Gamma_2 &= (+; -1, 0) & b_2 &= X_2 + X_3 \\
\Gamma_3 &= (\mu; 2) & b_3 &= \mu b_2 = \mu X_2 + \mu X_3 \\
\Gamma_4 &= (+; 1, 3) & b_4 &= b_1 + b_3 = \lambda X_1 + \mu X_2 + \mu X_3 \\
\Gamma_5 &= (*; 4, 2) & b_5 &= b_4 \cdot b_2 = (\lambda X_1 + \mu X_2 + \mu X_3)(X_2 + X_3) \;.
\end{aligned}
$$

Hence $B_j = e_j$ for $-2 \leq j \leq 0$ and

$$
\begin{aligned}
B_1 &= e_1 + \lambda B_{-2}, & B_2 &= e_2 + B_{-1} + B_0, & B_3 &= e_3 + \mu B_2, \\
B_4 &= e_4 + B_1 + B_3, & B_5 &= e_5 + b_2 \cdot B_4.
\end{aligned}
$$

This yields

$$B = \begin{pmatrix}
 & B_{-2} & B_{-1} & B_0 & B_1 & B_2 & B_3 & B_4 & B_5 \\
X_1 & 1 & & & & \lambda & & \lambda & \lambda b_2 \\
X_2 & & 1 & & & 1 & \mu & \mu & \mu b_2 \\
X_3 & & & 1 & & 1 & \mu & \mu & \mu b_2 \\
b_1 & & & & 1 & & & 1 & b_2 \\
b_2 & & & & & 1 & \mu & \mu & \mu b_2 \\
b_3 & & & & & & 1 & 1 & b_2 \\
b_4 & & & & & & & 1 & b_2 \\
b_5 & & & & & & & & 1
\end{pmatrix}$$

We continue with the proof and mention some properties of the matrix B:

(C) $B = (b_{ij})$ is an upper triangular matrix with $b_{ii} = 1$ for all $i \in N$. Furthermore, $b_{ij} \neq 0$ implies $i \leq_\Gamma j$.

The fact that B is an upper triangular matrix with ones on all diagonal positions follows by construction. We prove the second statement by induction on j. W.l.o.g. $j \geq 1$, $i < j$ and $b_{ij} \neq 0$. Suppose first that $j \in A$. Then $b_{ij} = b_{ij'} \pm b_{ij''} \neq 0$, hence $b_{ij'} \neq 0$ or $b_{ij''} \neq 0$. By induction we know that $i \leq_\Gamma j'$ or $i \leq_\Gamma j''$. Combining this with $j' \leq_\Gamma j$ and $j'' \leq_\Gamma j$ we get $i \leq_\Gamma j$. If $j \in M$, then $b_{ij} = b_{j''}b_{ij'} \neq 0$; hence $b_{ij'} \neq 0$. Again by induction one knows that $i \leq_\Gamma j'$, and as $j' \leq_\Gamma j$ we also have $i \leq_\Gamma j$. The case $j \in S$ is handled analogously. This completes the proof of (C).

(D) Each nonzero b_{ij} is a homogeneous polynomial of degree $d_j - d_i$.

We proceed by induction on j. W.l.o.g. $j \geq 1$, $i < j$, and $b_{ij} \neq 0$. If $j \in S$ then $b_{ij} = \lambda_j b_{ij'}$ and $d_j = d_{j'}$. By induction, $b_{ij'}$ is homogeneous of degree $d_{j'} - d_i = d_j - d_i$. Hence the same is true for b_{ij}. In the remaining cases we have $b_{ij} = b_{ij'} \pm b_{ij''}$ or $b_{ij} = b_{ij'} \cdot b_{j''}$. By induction we know that $b_{ij'}$ and $b_{ij''}$ are homogeneous or zero, and $b_{ij'} \neq 0$ implies $\deg b_{ij'} = d_{j'} - d_i$, and $b_{ij''} \neq 0$ implies $\deg b_{ij''} = d_{j''} - d_i$. If $j \in A$, then we have by (B) that $b_{j'} \cdot b_{j''} \neq 0$ and $d_{j'} = d_{j''} = d_j$; hence b_{ij} is homogeneous as well, and $\deg b_{ij} = d_j - d_i$. If $j \in M$, then b_{ij} is homogeneous of degree $\deg(b_{ij'}) + d_{j''} = d_{j'} - d_i + d_{j''} = d_j - d_i$. This proves (D).

Next we introduce $\leq_\Gamma$-antichains that will help to describe small depth computations. For every $a \geq 1$ we define

$$\Gamma_b(a) := \{t \in \underline{r} \mid t \in M; d_{t'}, d_{t''} \leq a < d_t\} \ .$$

For every a, $\Gamma_b(a)$ is a $\leq_\Gamma$-antichain that is involved in the computation of the b_i and b_{ij} in the following way.

(E) For $a \geq 1$ and $i, j \leq r$ with $d_i \leq a < d_j$ the following hold:

$$b_{ij} = \sum_{t \in \Gamma_b(a)} b_{it} b_{tj} \quad \text{and} \quad b_j = \sum_{t \in \Gamma_b(a)} b_t b_{tj} \ .$$

First note that $i \neq j$. As $d_j > a \geq 1$ we also have $j \geq 1$. Keeping $a, \Gamma,$ and i fixed, we prove the claims by induction on j. The start being clear, we focus on the induction step.

Case 1. $\Gamma_j = (\pm; j', j'')$.
On the one hand $b_j = b_{j'} \pm b_{j''}$ and $d_j = d_{j'} = d_{j''}$ and on the other hand $j', j'' < j$. Thus

$$b_{ij} = b_{ij'} \pm b_{ij''} = \sum_{t \in \Gamma_b(a)} b_{it}(b_{tj'} \pm b_{tj''}) = \sum_{t \in \Gamma_b(a)} b_{it}b_{tj} \ ,$$

and

$$b_j = b_{j'} \pm b_{j''} = \sum_{t \in \Gamma_b(a)} b_t(b_{tj'} \pm b_{tj''}) = \sum_{t \in \Gamma_b(a)} b_t b_{tj} \ .$$

Case 2. $\Gamma_j = (*; j', j'')$.
We distinguish two subcases.

Case 2.1. $d_{j'} \leq a$.
By (B) $d_{j''} \leq d_{j'}$, hence $j \in \Gamma_b(a)$. As $\Gamma_b(a)$ is an antichain, we have $b_{tj} = 0$ for all $t \in \Gamma_b(a) \setminus \{j\}$. Thus, as $b_{jj} = 1$,

$$b_{ij} = \underbrace{b_{ij}b_{jj}}_{t=j} + \sum_{t \in \Gamma_b(a) \setminus \{j\}} b_{it} \underbrace{b_{tj}}_{=0} \ , \quad \text{and} \quad b_j = \underbrace{b_j \, b_{jj}}_{t=j} + \sum_{t \in \Gamma_b(a) \setminus \{j\}} b_t \underbrace{b_{tj}}_{=0} \ .$$

Case 2.2. $a < d_{j'} \leq d_j$.
Then $B_j = e_j + b_{j''}B_{j'}$, in particular, $b_{tj} = b_{tj'} \cdot b_{j''}$, for all $t < j$. Applying the induction hypothesis to j' we obtain

$$b_{ij} = b_{ij'}b_{j''} = \left(\sum_{t \in \Gamma_b(a)} b_{it}b_{tj'} \right) b_{j''} = \sum_{t \in \Gamma_b(a)} b_{it}b_{tj} \ .$$

Our second claim in this subcase can be shown in a similar way. As the case $\Gamma_j = (\lambda_j; j')$ can be handled analogously, (E) follows.

(F) There exists a homogeneous straight-line program Γ' of length $r' = O(r^3)$ and depth $D' = O(\log(r) \cdot \log(d))$, which computes the homogeneous parts of f. (Recall that r denotes the length of Γ.)

W.l.o.g. all $X_1, \ldots, X_n$ occur in f. Thus $r \geq n - 1$. We are going to construct Γ' in $\lceil \log d \rceil$ stages. Each stage will contribute at most $2 + \lceil \log r \rceil$ to the depth of the final program Γ'. (Thus $D' \leq \lceil \log d \rceil (2 + \lceil \log r \rceil)$.)

Stage 0. Compute all b_j and b_{ij} of degree $\leq 2^0 = 1$.
All these b_j and b_{ij} are linear forms $\sum_{\nu=1}^n a_\nu X_\nu$ in n indeterminates or constants. Thus Stage 0 can be done in depth

$$1 + \lceil \log n \rceil \leq 1 + \lceil \log(r + 1) \rceil \leq 2 + \lceil \log r \rceil \ .$$

Stage $\delta + 1$. Compute all b_j and b_{ij}, whose degrees are in the interval $(2^\delta, 2^{\delta+1}]$.

(By (A) and (D), we are done after $\lceil \log d \rceil$ stages.) We first concentrate on the b_j. Let $2^{\delta+1} \geq d_j > 2^\delta =: a$. By (E) we have

(G) $\quad b_j = \sum_{t \in \Gamma_b(2^\delta)} b_t b_{tj} = \sum_{t \in \Gamma_b(2^\delta)} b_{t'} b_{t''} b_{tj}$

and by (D) and the definition of $\Gamma_b(2^\delta)$, the three polynomials $b_{t'}, b_{t''}$ and b_{tj} all have degree $\leq 2^\delta$, for every $t \in \Gamma_b(2^\delta)$. Hence these polynomials have already been computed in previous stages. Thus those b_j can be computed in additional depth at most

$$2 + \lceil \log(|\Gamma_b(2^\delta)|) \rceil \leq 2 + \lceil \log r \rceil \ .$$

Next we consider the b_{ij}. Let $2^{\delta+1} \geq \deg b_{ij} = d_j - d_i > 2^\delta$. Put $a := 2^\delta + d_i$. As $d_i \leq a < d_j$, we obtain by (E)

(H) $\quad b_{ij} = \sum_{t \in \Gamma_b(a)} b_{it} b_{tj} = \sum_{t \in \Gamma_b(2^\delta)} b_{t''} b_{it'} b_{tj} \ \ .$

Now both $b_{it'}$ and b_{tj} are of degree $\leq 2^\delta$, hence have been computed earlier. However, $b_{t''}$ might have a larger degree, say $d_{t'} \geq d_{t''} > 2^{\delta+1}$. We claim that in this case $b_{t''} b_{it'} b_{tj} = 0$. In fact, if $b_{it'} \neq 0$ and $d_{t''} > 2^{\delta+1}$, then $d_{t'} \geq d_i$, hence $d_t = d_{t'} + d_{t''} > d_i + 2^{\delta+1} \geq d_j$, thus $b_{tj} = 0$. Altogether, the b_{ij} in question can be computed with additional depth $\leq 2 + \lceil \log r \rceil$ and the resulting straight-line program Γ' satisfies the depth requirements stated in (F). Equations (G), (H), and the fact that $r \geq n - 1$ imply that $r' = O(r^3)$. This proves (F). Finally, combining Lemma 2.14 and (F) our claims follow. $\square$

2.5   Completeness of the Determinant Family

It is unknown whether the determinant family DET is VP-complete with respect to p-projections. We discuss here an interesting result, which states that DET is complete in a slightly bigger complexity class with respect to a more generous kind of reduction. Thm. 2.23 of the last section will be crucial for obtaining this.

A function $t \colon \mathbb{N} \to \mathbb{N}$ is called *quasi-polynomially bounded* (*qp*-bounded), if there exists a positive constant c such that $t(n) \leq n^{O(\log^c n)}$.

Definition 2.26 A p-family $f = (f_n)$ is said to be *qp-computable* iff the complexity $L(f_n)$ is a *qp*-bounded function of n. The complexity class VQP consists of all *qp*-computable families over k.

It is clear that VP is contained in VQP. (One can prove that this inclusion is strict, see Sect. 8.2.) By replacing the complexity L with the formula size E in the above definition, we get the complexity class VQP_e of *qp-expressible* *families*. It turns out that VQP_e and VQP actually coincide.

Corollary 2.27 *We have* $\mathrm{VQP}_e = \mathrm{VQP}$ *over any field.*

For the proof we need a lemma. Let $D(f)$ denote the *depth* of a polynomial f, i.e., the the minimal depth of a straight-line program computing f from variables and constants in k.

Lemma 2.28 *For any polynomial f we have* $\log(E(f) + 1) \leq D(f)$.

Proof. To a formula φ there corresponds a tree T_φ in a natural way. If $|T_\varphi|$ denotes its number of nodes, we have $E(\varphi) + 1 = |T_\varphi|$ and $\log|T_\varphi| \leq \mathrm{depth}(T_\varphi)$.

Let now Γ be a straight-line program of depth $D(f)$ which computes f. Such a straight-line program can be easily converted into a formula φ such that $\mathrm{depth}(T_\varphi) = \mathrm{depth}(\Gamma)$ and $\mathrm{val}(\varphi) = f$. (The size of φ might be exponential.) We then have $E(\varphi) + 1 = |T_\varphi| \leq 2^{D(f)}$, which proves the lemma. $\square$

We remark that this lemma is optimal in the sense that one can show that $D(f) = O(\log E(f))$ (Brent [15]).

Proof. (of Cor. 2.27) Take $(f_n) \in \mathrm{VQP}$. Then $n \mapsto L(f_n)$ is qp-bounded and $n \mapsto \deg f_n$ is p-bounded. Thm. 2.23 implies that $D(f_n) = O(\log^c n)$ for some constant c. From Lemma 2.28 we get that $n \mapsto E(f_n)$ is qp-bounded. Therefore, $(f_n) \in \mathrm{VQP}_e$. $\square$

Let $f = (f_n)$ and $g = (g_n)$ be p-families over k. In analogy to Def. 2.8, we say that f is a *qp-projection* of g iff there exists a qp-bounded function t, which is also p-bounded from below, such that f_n is a projection of $g_{t(n)}$ for all sufficiently large n. This defines a transitive relation.

Corollary 2.29 *The determinant family* DET *is VQP-complete with respect to qp-projections.*

For the proof we need the following result, which can be obtained in a similar way as Prop. 2.16 (compare also [21, Thm. 21.27]).

Proposition 2.30 *For any polynomial $f \in k[X_1, \ldots, X_n]$ of formula size e there is a square matrix A of size $e + 3$ over $k \cup \{X_1, \ldots, X_n\}$ such that $f = \det(A)$.*

Proof. (of Cor. 2.29) We know that DET $\in$ VP $\subseteq$ VQP. Let $f = (f_n) \in \mathrm{VQP}$. By Cor. 2.27 we know that $n \mapsto e_n = E(f_n)$ is qp-bounded. The above proposition tells us that f_n is a projection of DET_{e_n+3}. Thus f is a qp-projection of DET. $\square$

Valiant's hypothesis is equivalent to the statement PER $\notin$ VP. It is conjectured that even the stronger statement PER $\notin$ VQP is true. We call this *Valiant's extended hypothesis.*

Our last corollary gives different characterizations of this hypothesis. It is an immediate consequence of the VNP-completeness of the PER and the VQP-completeness of DET.

Corollary 2.31 *Over fields of characteristic different from two, the extended Valiant hypothesis is equivalent to each of the following statements:*

(1) $\text{VNP} \not\subseteq \text{VQP}$.

(2) $\text{PER} \notin \text{VQP}$.

(3) PER *is not a qp-projection of* DET.

It is remarkable that the last formulation is a purely algebraic one. It connects our theory to the classical problem of deriving the permanent from the determinant by substitution (compare Marcus and Minc [80], and von zur Gathen [42]).

We will prove in Sect. 8.2 that VQP is not contained in VNP.

3

Some Complete Families
of Polynomials

To a graph property $\mathcal{E}$ and a graph G we assign two polynomial functions in the edge weights of G: their generating and probability generating function. If the edge weights are interpreted as independent probabilities, then the first function gives the expected number of random spanning subgraphs G' of G having property $\mathcal{E}$, whereas the second one yields the probability that G' has property $\mathcal{E}$. Our goal is to study the complexity of the families of generating functions corresponding to certain sequences of graphs.

We report first about the few known nontrivial graph properties, which lead to p-computable generating functions. Then we provide completeness proofs for families of generating functions corresponding to graph properties such as matchings, cliques, graph factors, and connectivity.

3.1 Generating Functions of Graph Properties

Assume $\mathcal{E}$ is a *graph property*, that is, a class of (finite) graphs which contain with a graph also all its isomorphic copies. We want to study certain generating functions of the set of spanning subgraphs of a given graph which have property $\mathcal{E}$.

By an *edge weighted graph* we will understand a graph $G = (V, E)$ together with a weight function $w\colon E \to I := k \cup X$. Hereby, X denotes a finite set of indeterminates over a field k. We extend the weight function w to a function $w\colon 2^E \to k[X]$ by setting $w(E') := \prod_{e \in E'} w(e)$ for subsets E' of E. (The empty product is supposed to equal 1.)

Definition 3.1 The *generating function* $\mathrm{GF}(G, \mathcal{E})$ corresponding to an edge weighted graph G and a graph property $\mathcal{E}$ is defined as

$$\mathrm{GF}(G, \mathcal{E}) := \sum_{E' \subseteq E} w(E') \ ,$$

where the sum is over all subsets E' of E such that the spanning subgraph (V, E') of G has property $\mathcal{E}$.

To motivate this definition, assume for the moment that the edges of the graph G carry natural weights according to the map $a\colon E \to \mathbb{N}$. We define the additive weight $a(E')$ of a set of edges E' by $a(E') := \sum_{e \in E'} a(e)$. Let $N(\alpha)$ be the number of E' of additive weight α such that the spanning subgraph (V, E') has property $\mathcal{E}$. We introduce now the associated "multiplicative" weights $w(e)$ by setting $w(e) := T^{a(e)}$, where T is a new indeterminate. Then we have

$$(3.1) \qquad \mathrm{GF}(G, \mathcal{E}) = \sum_{\alpha \in \mathbb{N}} N(\alpha) T^{\alpha} \ .$$

In particular, the number of spanning subgraphs of G having property $\mathcal{E}$ is obtained by evaluating $\mathrm{GF}(G, \mathcal{E})$ for $T = 1$. The coefficients $N(\alpha)$ can be obtained by evaluating $\mathrm{GF}(G, \mathcal{E})$ for several values of T by interpolation.

The generating function $\mathrm{GF}(G, \mathcal{E})$ can also be given the following natural interpretation. Generate a spanning subgraph of G by picking independently each edge e of G at random with probability $w(e) \in [0, 1]$. Then $\mathrm{GF}(G, \mathcal{E})$ is the expected number of spanning subgraphs of G which have property $\mathcal{E}$.

Let K_n denote the complete graph on n nodes and let $K_{n,n}$ be the complete bipartite graph with two sets of nodes of size n. Unless otherwise stated, we will always assume that these graphs are endowed with a weight function mapping their edges to independent indeterminates. We remark that $\mathrm{GF}(K_n, \mathcal{E})$ is universal in the following sense: for any edge weighted graph G on n nodes, the generating function $\mathrm{GF}(G, \mathcal{E})$ is a projection of $\mathrm{GF}(K_n, \mathcal{E})$. The polynomial $\mathrm{GF}(K_{n,n}, \mathcal{E})$ shares a similar universal property for the class of bipartite graphs.

The purpose of our investigations is to find out for which graph properties $\mathcal{E}$ the corresponding family of generating functions $\mathrm{GF}(K_n, \mathcal{E})$ or $\mathrm{GF}(K_{n,n}, \mathcal{E})$ is VNP-complete. We will also ask this question for the sequences of *rectangular lattice graphs* R_n and *cubic lattice graphs* C_n. The graph R_n has the vertex set $\{(i, j) \mid 1 \leq i, j \leq n\}$ with integer coordinates and its edges are the pairs of nodes separated by (Euclidean) unit distance. The cubic lattice graph C_n has the vertex set $\{(i, j, \ell) \mid 1 \leq i, j \leq n, 0 \leq \ell \leq 1\}$, where again the edges consist of the pairs of nodes separated by unit distance. We note the following obvious relations ($\leq$ denotes the projection)

$$\mathrm{GF}(R_n, \mathcal{E}) \leq \mathrm{GF}(C_n, \mathcal{E}) \leq \mathrm{GF}(K_{n^2, n^2}, \mathcal{E}) \leq \mathrm{GF}(K_{2n^2}, \mathcal{E}) \ .$$

In the sequel, we will understand by a *p-sequence of graphs* a sequence of graphs (G_n) such that the number of nodes of G_n is p-bounded in n.

From Valiant's criterion 2.20 we easily get the following proposition, which shows the p-definability of all families of generating functions we will be studying in the sequel. (For a definition of the nonuniform complexity class P/poly see Sect. 4.3.)

Proposition 3.2 *Assume that the problem to test whether a given graph has property $\mathcal{E}$ is in* P/poly. *Then the sequence of generating functions* $(\mathrm{GF}(K_n, \mathcal{E}))$ *is p-definable over any field.*

Sometimes, it is more natural to consider a different generating function. We define the *probability generating function* $\mathrm{PGF}(G, \mathcal{E})$ corresponding to G and $\mathcal{E}$ as follows:

$$\mathrm{PGF}(G, \mathcal{E}) := \sum_{E' \subseteq E} \prod_{e \in E'} w(e) \prod_{e \in E \setminus E'} (1 - w(e)) \; ,$$

where the sum is over all subsets E' of E such that the spanning subgraph (V, E') of G has property $\mathcal{E}$. This function has the following natural interpretation. Generate a spanning subgraph of G by picking independently each edge e of G at random with probability $w(e) \in [0, 1]$. Then $\mathrm{PGF}(G, \mathcal{E})$ is the probability that the resulting subgraph has property $\mathcal{E}$. The function PGF is of importance for determining the reliability of communication networks (cf. Sect. 3.3.7).

We further remark that Valiant's criterion 2.20 implies as well that the families $(\mathrm{PGF}(K_n, \mathcal{E}))$ are p-definable, for graph properties $\mathcal{E}$ which can be tested in nonuniform polynomial time on a Turing machine.

The generating functions GF and PGF are related as follows. Suppose $G = (V, E)$ is an edge weighted graph with respect to $w \colon E \to I$. Let G^* stand for the graph G together with the function $w^* \colon E \to k[X]$, $w^*(e) := w(e)/(1 - w(e))$. (This is strictly speaking not a weight function in our usual sense.) Then we have

$$\mathrm{PGF}(G, \mathcal{E}) = \mathrm{GF}(G^*, \mathcal{E}) \; \textstyle\prod_{e \in E}(1 - w(e)) \; .$$

This relation shows that the complexities of $\mathrm{PGF}(K_n, \mathcal{E})$ and $\mathrm{GF}(K_n, \mathcal{E})$ are polynomially related. However, this reduction involves divisions. Using a similar idea as in Strassen [104], one can show that the use of divisions can be avoided. This way, one can prove that the families $(\mathrm{PGF}(K_n, \mathcal{E}))$ and $(\mathrm{GF}(K_n, \mathcal{E}))$ are equivalent with respect to the c-reduction, which will introduced in Sect. 5.4 (over infinite fields). Hence for proving VNP-completeness with respect to c-reduction, it is the same whether we work with GF or PGF.

3.2   p-Computable Families

We report here about the few known nontrivial graph properties which lead to p-computable generating functions.

To warm up, consider first the trivial graph property $\mathcal{E}_{\mathrm{triv}}$ fulfilled by all graphs. Let E denote the set of edges of K_n. Then we may write

$$\mathrm{GF}(K_n, \mathcal{E}_{\mathrm{triv}}) = \sum_{E' \subseteq E} \prod_{e \in E'} X_e = \prod_{e \in E}(1 + X_e) \; ,$$

which shows that $(\mathrm{GF}(K_n, \mathcal{E}_{\mathrm{triv}}))$ is p-computable.

Now consider the graph property consisting of all trees. Let X_{ij} be indeterminates for $1 \le i, j \le n$ and put $X_{ji} = X_{ij}$. Define the n by n matrix $A = [a_{ij}]$ by setting $a_{ij} := -X_{ij}$ for $i \ne j$ and $a_{ii} := -\sum_{\ell \ne i} X_{i\ell}$. Let B be a matrix arising from A by deleting one row and one column of the same index. Then a well-known result due to Kirchhoff (cf. [14, p. 40]) states that

$$\mathrm{GF}(K_n, \{\text{trees}\}) = \det(B) \ .$$

This immediately implies the following.

Theorem 3.3 *The family* $(\mathrm{GF}(K_n, \{\text{trees}\}))$ *is p-computable.*

Now let $\mathcal{DI}$ be the graph property expressing that all connected components have exactly two nodes. The corresponding generating function enumerates the *perfect matchings*, or *dimer coverings*. For a detailed treatment of the following with a discussion of applications to crystal physics, see Kasteleyn [64].

Suppose $G = (\underline{n}, E)$ is a given weighted graph and let $X_{ij} = X_{ji}$ be the weight of the edge $\{i, j\} \in E$. In the sequel, we may assume that n is even. We orient each edge of G in a certain way, which will be specified later. We then consider the skew-symmetric n by n matrix $A = [a_{ij}]$ defined by $a_{ij} := X_{ij}$ if the edge (i, j) has positive orientation and $a_{ij} := -X_{ij}$ otherwise. It is well-known that the determinant of a skew-symmetric matrix A is the square of its so-called *Pfaffian* $\mathrm{Pf}(A)$. This function can be explicitly described as

$$\mathrm{Pf}(A) = \sum_{\sigma} \mathrm{sgn}(\sigma)\, a_{\sigma(1)\sigma(2)} a_{\sigma(3)\sigma(4)} \cdots a_{\sigma(n-1)\sigma(n)} \ ,$$

where the sum is over all permutations $\sigma \in S_n$ such that $\sigma(1) < \sigma(2)$, $\sigma(3) < \sigma(4), \ldots, \sigma(n-1) < \sigma(n)$, and $\sigma(1) < \sigma(3) < \ldots < \sigma(n-1)$. (Cf. [64, p. 87].) This implies that

$$\mathrm{Pf}(A) = \sum_{E'} \epsilon_{E'} \prod_{e \in E'} X_e \ ,$$

where the sum runs over all perfect matchings E' of G and $\epsilon_{E'} \in \{-1, +1\}$ is a sign depending on E'. The point is now that for planar graphs G, the orientation of the edges can be chosen such that the map $E' \mapsto \epsilon_{E'}$ is constant, say equal to one (consistent orientation). Assuming this, we therefore have $\mathrm{GF}(G, \mathcal{DI}) = \mathrm{Pf}(A)$. (This discovery is due to Fisher [36] and Kasteleyn [63].)

The Pfaffian of an n by n skew-symmetric matrix can be computed with $O(n^3)$ arithmetic operations by a variant of Gaussian elimination, based on the fact that $\mathrm{Pf}(SAS^{\top}) = \det(S)\mathrm{Pf}(A)$. (In fact, one can even prove that the "exponent" of this problem is the same as that of matrix multiplication, cf. [21, Chap. 16].)

Putting all this together we obtain the following.

Theorem 3.4 *The family* $(\mathrm{GF}(G_n, \mathcal{DI}))$ *is p-computable for any p-sequence* (G_n) *of planar graphs.*

A graph is called *closed* if all of its nodes have even degree. Let $\mathcal{IS}$ be the set of closed graphs. The generating function $\mathrm{GF}(R_n, \mathcal{IS})$ naturally occurs in the partition function of the so-called *Ising problem*, a famous problem in statistical physics. (For details see Kastèleyn [64].) Similarly as in the dimer covering problem, one can express the generating function $\mathrm{GF}(R_n, \mathcal{IS})$ as a certain Pfaffian and obtains the following.

Theorem 3.5 *The family $(\mathrm{GF}(R_n, \mathcal{IS}))$ is p-computable.*

We remark that Courcelle et al. [27] have recently obtained a general result on the p-computability of the generating functions of certain logically characterized graph properties when restricting attention to graphs of *bounded tree width*. In particular, the permanent and Hamilton cycle polynomial can be evaluated in a polynomial number of arithmetic operations at matrices of bounded tree width. We would also like to mention Barvinok's result [6] stating that the permanent and the Hamilton cycle polynomial can evaluated at matrices of *bounded rank* in a polynomial number of arithmetic operations.

The fast algorithms for the generating functions considered in this section are all based on algebraic properties of certain well-behaved polynomial functions, namely determinants and Pfaffians. This suggests that our algebraic model of computation is very adequate for studying such problems.

In this book we will focus on the complexity of exact algebraic computations. We remark that when allowing for *approximate solutions* and *randomization*, a wealth of other interesting results has been obtained for counting problems: for an overview and references the reader may consult the book by Motwani and Raghavan [86]. Of particular importance is the randomized approximation algorithm for the permanent due to Jerrum and Sinclair [54]. (See also Sinclair [99].) A general survey on counting algorithms can be found in Welsh [118].

3.3 VNP-Complete Families

We first report on known results due to Valiant and Jerrum, which identify certain graph properties having VNP-complete families of generating functions. Then we present some new results of ours of this type.

It is well-known that the familiar reductions used in NP-completeness proofs can often be modified to yield parsimonious reductions, thus giving proofs of #P-completeness. In turns out that variants of such reductions can sometimes be used to obtain p-projections, thus showing the VNP-completeness of the corresponding generating functions.

Unless otherwise stated, we shall mean by the completeness of a p-family always VNP-completeness with respect to p-projection. Throughout this section we work over a field k of characteristic different from two.

3.3.1   Matchings

The generating function of perfect matchings in complete bipartite graphs is the permanent polynomial: $\mathrm{GF}(K_{n,n}, \mathcal{DI}) = \mathrm{PER}_n$.

All attempts to generalize the technique of Kasteleyn and Fisher (compare Thm. 3.4) for the generating function of dimer coverings to nonplanar graphs have failed. The key result in Valiant [107] gives an explanation of this fact from the point of view of computational complexity. Recall that C_n denotes a cubic lattice graph.

Theorem 3.6 (Valiant) *The family* $(\mathrm{GF}(C_n, \mathcal{DI}))$ *is complete.*

Let $\mathcal{MD}$ be the graph property expressing that all connected components have at most two nodes. The corresponding generating function enumerates the *partial matchings* or *monomer-dimer coverings*. For bipartite graphs, we have

$$\mathrm{PER}_n^* := \mathrm{GF}(K_{n,n}, \mathcal{MD}) = \sum_{\pi} \prod_{i \in \mathrm{def}\,\pi} X_{i\pi(i)} \ ,$$

where the sum is over all injective partial maps $\underline{n} \to \underline{n}$. (The zero product is assumed to be 1.) We call PER_n^* the *partial permanent* and write PER^* for the corresponding p-family.

Theorem 3.7 (Jerrum) *The family* PER^* *of partial permanents is complete.*

Proof. Let $\{u_1, \ldots, u_n\}$ and $\{v_1, \ldots, v_n\}$ be the two classes of nodes of $K_{n,n}$. For $i \in \underline{n}$ we introduce additional nodes u_i', v_i' and edges $\{u_i', v_i\}$, $\{u_i, v_i'\}$ of weight -1. (The weight of $\{u_i, v_j\}$ is assumed to be the indeterminate X_{ij}, as usual.) We denote the resulting weighted bipartite graph by G.

We claim that $\mathrm{GF}(K_{n,n}, \mathcal{DI}) = \mathrm{GF}(G, \mathcal{MD})$. This implies that PER_n is a projection of PER_{2n}^*. Hence PER is a p-projection of PER^* and Thm. 2.10 yields the assertion.

To prove the claim, assume that π is a partial matching of $K_{n,n}$ with free nodes u_i, v_j for $i \in I(\pi)$ and $j \in J(\pi)$. Choose subsets $M \subseteq I(\pi)$, $N \subseteq J(\pi)$ and add the edges $\{u_i, v_i'\}$ for $i \in M$ and $\{u_j', v_j\}$ for $j \in N$. Note that the resulting partial matching of G has the weight $(-1)^{|M|+|N|} w(\pi)$. Moreover, all partial matchings of G can be obtained from a uniquely determined π and subsets M, N in this way. Therefore,

$$\mathrm{GF}(G, \mathcal{MD}) = \sum_{\pi} w(\pi) \sum_{M \subseteq I(\pi)} (-1)^{|M|} \sum_{N \subseteq J(\pi)} (-1)^{|N|} \ .$$

However, the sum $\sum_{M \subseteq I(\pi)} (-1)^{|M|}$ vanishes unless $I(\pi)$ is empty. Hence only perfect matchings contribute to the above sum and the claim follows. $\qquad\square$

It turns out that the family PER* is a flexible tool for proving VNP-completeness, similarly as the problem 3-SAT in the theory of NP-completeness. We think it would be worthwile to solve the following problem, since a positive answer to this would allow to remove the assumption "char$k \neq 2$" in many of the following statements on VNP-completeness.

Problem 3.1 Is PER* also VNP-complete over fields of characteristic two?

Of course, this problem is related with the question whether computing the parity of the number of partial matchings in bipartite graphs is a $\oplus$P-complete problem. Note that PER is p-computable over fields of characteristic two, since this family then coincides with the determinant family.

The following result due to Jerrum [52] explains why the technique of Fisher and Kasteleyn for the generating functions of dimer coverings in planar lattices cannot be generalized to monomer-dimer coverings. Recall that R_n denotes a rectangular lattice graph.

Theorem 3.8 (Jerrum) *The family* $(\mathrm{GF}(R_n, \mathcal{MD}))$ *is complete.*

With respect to the proof some comments are in order. The reduction in Jerrum's thesis [52] contained as constants nonrational algebraic numbers and proved the above theorem over some number field (or the reals). In his later article [53], a proof for the #P-completeness to enumerate monomer-dimer coverings in planar graphs was given. By combining the simplified construction for simulating crossovers in this article with the approach in Jerrum's thesis, a proof of the above result can be obtained.

Jerrum [52] has also explained why the techniques for solving the 2-dimensional Ising problem are doomed to fail in dimension three.

Theorem 3.9 (Jerrum) *The family* $(\mathrm{GF}(C_n, \mathcal{IS}))$ *is complete over the reals.*

3.3.2 Cliques

A *clique* in a graph $G = (V, E)$ is a nonempty subset $U \subseteq V$ such that all pairs of nodes in U are connected by an edge. We denote the set of these edges by E_U. Let $\mathcal{CL}$ be the set of graphs, where one connected component is a complete graph and the remaining connected components consist of one node only. The generating function $\mathrm{GF}(G, \mathcal{CL})$ enumerates the cliques in a graph G.

Theorem 3.10 *The family* $(\mathrm{GF}(K_n, \mathcal{CL}))$ *is complete.*

Proof. Consider the graph $G = (V, E)$, where $V = \{0, 1, \ldots, n\}^2$, and where two nodes (i, j) and (i', j') are connected iff $i \neq i'$ and $j \neq j'$. Observe that the cliques in G can be identified with the nonempty partial matchings of $K_{n,n}$. We define the weight $w(e)$ of an edge $e = \{(i,j),(i',j')\}$ to be X_{ij} if $(i',j') = (0,0)$ and 1 otherwise. If a clique U of G does not contain $(0,0)$, then $w(E_U) = 1$. Otherwise, if $U = W \cup \{(0,0)\}$, we have $w(E_U) = \prod_{(i,j)\in W} X_{ij}$. This implies that

$$\mathrm{GF}(G,\mathcal{CL}) = \mathrm{PER}_n^* + N_n \ ,$$

where N_n is the number of cliques in G not containing the node $(0,0)$. Now form the disjoint union of G with the graph K_2 , whose single edge carries the weight $1 - N_n$. Then we get for the resulting weighted graph G', that $\mathrm{GF}(G',\mathcal{CL}) = \mathrm{GF}(G,\mathcal{CL}) - N_n = \mathrm{PER}_n^*$. The assertion follows now with Thm. 3.7. $\qquad\square$

3.3.3   Cycle Format Polynomials

We may write a partition of $n \in \mathbb{N}$ in frequency notation $\rho = (\rho(1),\rho(2),\dots)$, where $\rho(i)$ denotes the number of pieces of size i. We note that always $\sum i\rho(i) = n$. It is convenient to write such a partition symbolically as $\rho = 1^{\rho(1)}2^{\rho(2)}\cdots n^{\rho(n)}$, or $\rho \models n$. We associate with the cycle format ρ the graph property $\mathcal{CF}_\rho$ describing all graphs on n nodes consisting of $\rho(i)$ disjoint i-cycles, for $i = 1,2,\dots n$. (The graphs K_1 and K_2 are to be considered as 1-cycles and 2-cycles, respectively.) The corresponding generating function

$$\mathrm{CF}_\rho := \mathrm{GF}(K_n,\mathcal{CF}_\rho)$$

will be called the *cycle format polynomial* of ρ.

Two particular cases deserve special attention. The cycle format polynomial of the partition $\rho = 2^{n/2}$ is just the generating function $\mathrm{GF}(K_n,\mathcal{DI})$ of perfect matchings (n even). The cycle format polynomial of the partition $\rho = n^1$ is the generating function corresponding to the graph property consisting of the cycles. We call it the *undirected Hamilton cycle polynomial* UHC_n, since it enumerates the undirected Hamilton cycles (i.e., spanning cycles) in a graph. This should not be confused with the *directed* Hamilton cycle polynomial HC_n which enumerates the directed Hamilton cycles in a digraph. We leave it to the reader to check that UHC_m is a projection of UHC_n if $n \geq m + 2$.

By Thm. 2.10 we know already that the families $(\mathrm{GF}(K_n,\mathcal{DI}))$ and (HC_n) are complete. As a byproduct of our later investigations in Sect. 3.3.5, we will obtain that the family (UHC_n) of undirected Hamilton cycle polynomials is complete as well (see Cor. 3.19). We will use this corollary in the sequel.

Our goal is to extend these results as follows.

Theorem 3.11 *Let $\rho_n \models n$ be a sequence of partitions such that there exists some $\epsilon > 0$ with $n - \rho_n(1) \geq n^\epsilon$ for all n. Then the corresponding sequence (CF_{ρ_n}) of cycle format polynomials is complete.*

Before giving the proof, we present some auxiliary results.

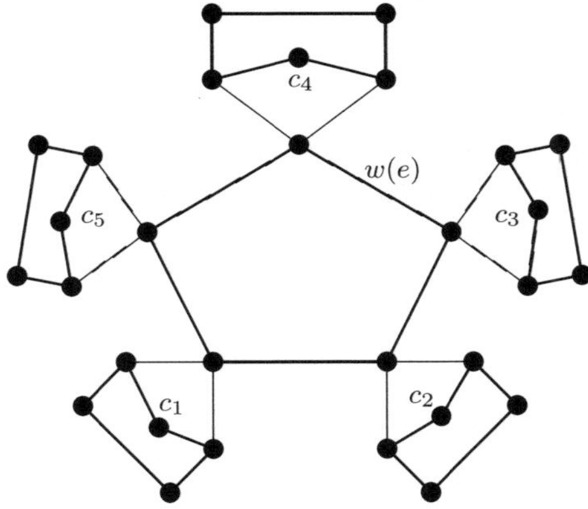

Figure 3.1: The graph S_5 and its covering by 5-cycles.

Lemma 3.12 *Let σ and ρ be partitions in frequency notation such that $\sigma(i) \leq \rho(i)$ for all i. Then CF_σ is a projection of CF_ρ.*

Proof. Assume $\sigma \models m$ and $\rho \models n$. Let G be the disjoint union of the complete graph K_m and $\rho(i) - \sigma(i)$ disjoint cycles of length i, for $i = 1, 2, \ldots$. All cycles are assumed to carry the weight 1. Then it is easy to check that $\mathrm{CF}_\sigma = \mathrm{GF}(G, \mathcal{CF}_\rho)$. $\square$

Proposition 3.13 PER_n *is a projection of* CF_{m^r} *if* $m \geq 2$ *and* $r \geq n + mn^2$.

Proof. Our construction is related to the one in Kirkpatrick and Hell [65]. We may w.l.o.g. assume that $m \geq 3$.

Take an m-cycle and duplicate one of its nodes including its adjacent edges. Let us denote the resulting graph by D_m and call the duplicated nodes the inner and outer connector, respectively. Now take m disjoint copies of D_m and join their inner connectors to form an m-cycle. Call the resulting auxiliary graph S_m and denote its outer connectors by $c_1, \ldots, c_m$ (see Fig. 3.1). The graph S_m has a unique covering C_+ by m-cycles. If we remove all the connectors (including their adjacent edges), then the resulting graph also has a unique covering by m-cycles, called C_-. The following *switching property* of S_m is crucial for us. Think of a disjoint union of S_m with a graph G_0 and identify the outer connectors of S_m with nodes of G_0. Then any covering by m-cycles of the resulting graph "covers" S_m by either a cycle cover C_+ or C_-.

Let $H = (V, R)$ be an m-hypergraph on a set V of mn nodes. By a perfect matching of H we understand a subset of its hyperedges which form a partition of V. We assign to H a graph G as follows. Start with the graph $(V, \emptyset)$. For each hyperedge $\{v_1, \ldots, v_m\}$ of H we insert a copy of our auxiliary graph S_m by identifying the nodes v_i with the outer connectors of S_m. (The sets of inner nodes of distinct copies of S_m are assumed to be disjoint.) The resulting graph G has exactly $|V| + |R|m^2$ nodes.

The point of our construction is the fact that there is a bijection between the perfect matchings of H and the coverings of G by m-cycles. (This follows readily from the switching property of S_m.) Assume now additionally that H comes with a weight function $w: R \to k[X]$. Define the weight of all edges of G to be 1 except that for each hyperedge e of G, we suppose exactly one of the edges of the "middle m-cycle" of S_m to carry the weight $w(e)$ (cf. Fig. 3.1). Then we have for $\rho = m^r$, $r = n + |R|m$ that

$$\mathrm{GF}(G, \mathcal{CF}_\rho) = \sum_\Pi \prod_{e \in \Pi} w(e) \ ,$$

where the sum is over all perfect matchings Π of H.

We specialize now the hypergraph $H = (V, R)$ as follows. Let V be the disjoint union of m copies of $\{1, 2, \ldots, n\}$ and let the hyperedges by given by the m-tuples $(i, j, \ldots, j)$ carrying the weight X_{ij} $(1 \le i, j \le n)$. Note that $|R| = n^2$. Then the perfect matchings of H correspond bijectively to the perfect matchings of the complete bipartite graph $K_{n,n}$ and we obtain

$$\mathrm{PER}_n = \sum_\Pi \prod_{e \in \Pi} w(e) \ .$$

Therefore, we arrive at $\mathrm{PER}_n = \mathrm{GF}(G, \mathcal{CF}_\rho)$, where $\rho = m^r$ and $r = n + mn^2$, which was to be shown. $\qquad\square$

Proof. (of Thm. 3.11) By our assumption $n - \rho_n(1) \ge n^\epsilon$ we may assume w.l.o.g. that $\rho_n(1) = 0$ for all n. We put

$$t_n := \max\{i \mid \rho_n(i) > 0\}, \quad I := \{n \mid t_{(n+2)^4} < n + 2\} \ .$$

Let h be the mixture of the families of permanents and undirected Hamilton cycle polynomials w.r.t. the index set I. By Thm. 2.10, Cor. 3.19, and Lemma 2.12 the family h is complete. We are going to show that h is a p-projection of the family (CF_{ρ_n}) of cycle format polynomials. We write $N := (n + 2)^4$.

Assume first that $n \notin I$, i.e., $t_N \ge n + 2$. By Lemma 3.12, UHC_{t_N} is a projection of CF_{ρ_N}. By monotonicity, we conclude that $h_n = \mathrm{UHC}_n$ is a projection of CF_{ρ_N}.

Assume now $n \in I$. Define r and m by $r := \rho_N(m) := \max_i \rho_N(i)$. Note that $2 \le m \le t_N \le n + 1$. From $N = \sum i\rho_N(i) \le r t_N$ we conclude that $r \ge (n + 2)^3$, hence $n + mn^2 \le r$. Prop. 3.13 implies that $h_n = \mathrm{PER}_n$ is a projection of CF_{m^r} and thus of CF_{ρ_N}. $\qquad\square$

Certain families of cycle format polynomials provide natural candidates of p-definable families, which are neither p-computable nor p-complete. To explain this, we first present an upper bound on the complexity of the undirected

Hamilton cycle polynomials, which easily follows by well-known dynamic programming techniques.

Proposition 3.14 *The total complexity of* UHC_n *satisfies* $L(\mathrm{UHC}_n) \leq n^2 2^n$.

Proof. (Compare Papadimitriou and Steiglitz [88, p. 450].) For a set $S \subseteq \underline{n}$ containing the distinct nodes 1 and m let $F(S, m)$ be the sum of the weights of all paths in K_n from 1 to m which visit all nodes in S exactly once. Obviously, $F(\{1, m\}, m) = X_{1m}$. We can compute the polynomials $F(S, m)$ by the following recursion formula valid for $|S| > 2$:

$$F(S, m) = \sum_{\mu \in S \setminus \{1, m\}} F(S \setminus \{m\}, \mu) \, X_{\mu m} \ .$$

In this way, we can compute all $F(S, m)$ with

$$\sum_{\sigma = 3}^{n} 2\sigma^2 \binom{n-1}{\sigma - 1} \leq 2n^2(2^{n-1} - n) = n^2 2^n - 2n^3$$

arithmetic operations. Taking into account that $\mathrm{UHC}_n = \sum_{m=2}^{n} F(\underline{n}, m) X_{m1}$, the assertion follows. $\qquad\square$

Valiant's hypothesis VP $\neq$ VNP states that complete families (f_n) of polynomials are not p-computable. The extended Valiant hypothesis (cf. Sect. 2.5) claims that the complexity growth of such polynomials f_n is not quasi-polynomially bounded. The following hypothesis is even stronger and claims that complete families have an exponential growth of complexity.

(H)   For any complete family (f_n) there is some $\epsilon > 0$ such that $L(f_n) \geq 2^{n^\epsilon}$ for infinitely many n.

It is easy to see that (H) is equivalent to the following statement: There is some p-definable family (g_n) and some $\epsilon > 0$ such that $L(g_n) \geq 2^{n^\epsilon}$ for infinitely many n.

Proposition 3.15 *Consider the family* (CF_{ρ_n}) *of cycle format polynomials, where* $\rho_n = 1^{n - t_n} \cdot t_n^1$ *and* $t_n := \lceil \log^c n \rceil$, $c > 0$. *Under the hypothesis (H), this family is neither complete nor p-computable, if c is chosen sufficiently large.*

Proof. We have by Prop. 3.14

$$L(\mathrm{CF}_{\rho_n}) \leq \binom{n}{t_n} L(\mathrm{UHC}_{t_n}) = 2^{O(t_n \log n)} = 2^{O(\log^{c+1} n)} \ .$$

Hence the complexity growth of CF_{ρ_n} is at most quasi-polynomial. Therefore, this family cannot be complete under hypothesis (H).

By the hypothesis (H) there is some $\epsilon > 0$ and a strictly increasing sequence (τ_i) such that $L(\mathrm{UHC}_{\tau_i}) \geq 2^{\tau_i^\epsilon}$ for all i. On the other hand, we have $L(\mathrm{UHC}_{t_n}) \leq L(\mathrm{CF}_{\rho_n})$ by Lemma 3.12. Now choose $n = n_i$ minimal such that $\tau_i + 2 \leq t_n$. Thus $\log^c(n_i - 1) \leq t_{n_i - 1} < \tau_i + 2$. Using the monotonicity of UHC_t we conclude that $2^{\tau_i^\epsilon} \leq L(\mathrm{CF}_{\rho_{n_i}})$. Thus (CF_{ρ_n}) is not p-computable if $c\epsilon > 1$. $\hfill\square$

For statements related to this proposition, which rely on weaker hypotheses, we refer to Chap. 5.

3.3.4 Graph Factors

In the sequel let F denote a connected graph. Let the graph property $\mathcal{FA}(F)$ describe the graphs all of whose connected components are isomorphic to F. A spanning subgraph of a graph G which has the property $\mathcal{FA}(F)$ will be called an F-factor of G. The corresponding generating functions

$$\mathrm{Fact}_n(F) := \mathrm{GF}(K_n, \mathcal{FA}(F))$$

will be called the *F-factor polynomials*. We remark that this notion contains certain cycle format polynomials as special cases. Namely, $\mathrm{CF}_\rho = \mathrm{Fact}_n(F)$ if F is an m-cycle and $\rho = m^r \models n$. Also note that $\mathrm{Fact}_n(K_1) = 1$.

Kirkpatrick and Hell [65] have proved that deciding the existence of an F-factor in a given graph is NP-complete if F has at least three nodes. With the same method we obtain the following nice result.

Theorem 3.16 *The family* $(\mathrm{Fact}_n(F))$ *is complete if F has at least two nodes.*

Proof. We proceed similarly as in the proof of Prop. 3.13. Let m denote the number of nodes of F. We duplicate a node v of F including its adjacent edges and denote the duplicated nodes by v_- and v_+, respectively. The resulting graph D contains two distinguished subgraphs F_- and F_+ isomorphic to F. (E.g., F_+ contains v_+ but not v_-.)

We say that D fulfilles the *switching property* if the following is true. Think of the disjoint union of D with any graph G_0 and identify the nodes v_- and v_+ of D with distinct nodes of G_0. Then any F-factor Φ of the resulting graph G_1 "covers" D by either F_- or F_+, that is, Φ contains either F_- or F_+.

Lemma 3.17 *There is always a node v of F such that the resulting graph D satisfies the switching property.*

A proof of this lemma will be given at the end. We assume in the sequel that v is chosen according to this lemma.

We shall construct a new graph S, called *star*, as follows. We start with a copy of F called the *body*. For each node u of the body we create a disjoint

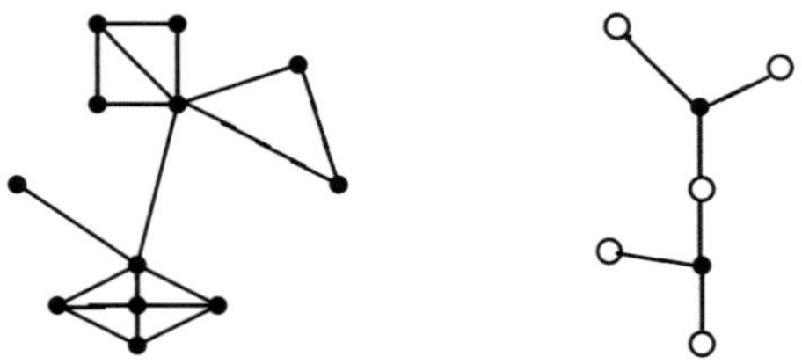

Figure 3.2: The graph F and its block-cutnode tree $bc(F)$.

copy of the graph D and identify u with the node v_- of this D. The m copies of D attached to the body this way will be called the *arms* of S, and the m nodes of type v_+ of the arms are said to be the *connectors* of S. Note that the star S is a graph on m^2 nodes. (Compare Fig. 3.1.)

The switching property of D implies the follows switching property of the star S. Think of the disjoint union of S with any graph G_0 and identify the connectors of S with distinct nodes of G_0. Then any F-factor of the resulting graph G_1 covers each arm of S either by a copy of F_+ or by a copy of F_-. It is immediate that only two possibilities exist: either all arms are covered by F_+ or all arms are covered by F_-.

Now let $H = (V, R)$ be an m-hypergraph on a set V of mn nodes. We assign to H a graph G as follows. We start with the graph $(V, \emptyset)$. For each hyperedge $\{v_1, \ldots, v_m\}$ of H we insert a copy of the star S by identifying the nodes v_i with the connectors of S. (The sets of inner nodes of distinct copies of S_m are assumed to be disjoint.) The resulting graph G has exactly $|V| + |R| m^2$ nodes.

By the switching property of S there is a bijection between the perfect matchings of H and the F-factors of G. Arguing exactly as in the proof of Prop. 3.13 we can get $\mathrm{PER}_n = \mathrm{GF}(G, \mathcal{FA}(F))$ for a suitable choice of the weight function of G. $\qquad\square$

Proof. (of Lemma 3.17) For the following facts from graph theory see [14, III.2]. A *cutnode* of F is a node whose deletion increases the number of connected components. A maximal connected subgraph of G without cutnode is called a *block* of F. Consider the *block-cutnode graph* $bc(F)$ whose vertices are the blocks and cutnodes of F, and whose edges join cutnodes to blocks containing it. One can show that $bc(F)$ is a tree (see Fig. 3.2). The leaves of this tree are the *endblocks* of F. We claim that if the node v is not a cutnode and if it is chosen in an endblock B_e of F, then the resulting graph D satisfies the switching property. Indeed, assume D to be built in a graph G_1 via the connecting nodes v_- and v_+ as usual and think of an F-factor Φ of G_1. By taking into account the tree structure of $bc(F)$, it is easy to see that all blocks of F with the possible exception of B_e have to be covered by the same component F' of Φ. Either v_- or v_+ has to belong to F', say $v_+ \in F'$. If F' were

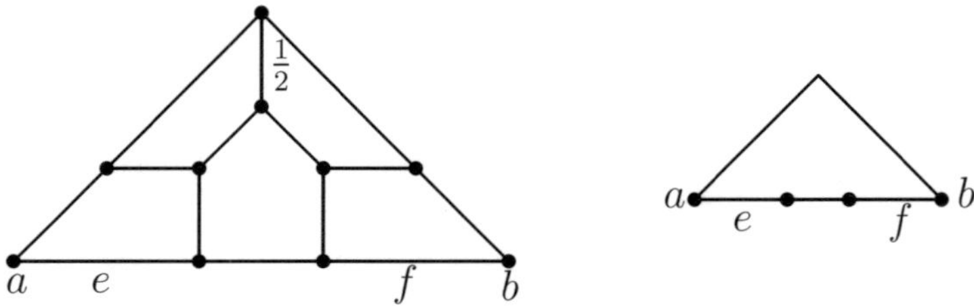

Figure 3.3: The auxiliary graph C and its symbol.

different from the copy F_+, then v_+ would be a necessarily a cutnode of F'. Hence the cutnodes of F would form a proper subset of the set of cutnodes of F', which is impossible, as F' is isomorphic to F. Therefore, $F' = F_+$ and the lemma is proved. $\qquad\square$

3.3.5   Hamilton Cycles of Planar Graphs

Garey, Johnson, and Tarjan [39] proved that deciding whether a planar, cubic graph contains a Hamilton cycle is NP-complete, by reducing 3-SAT to this problem. With the same overall strategy, but using simpler components, we can prove a corresponding result in the algebraic setting. Our simplifications come from the fact that we basically establish a reduction from the counting version of 2-SAT to the problem. More specifically, we reduce the partial permanent PER_n^* to the polynomials under consideration via a p-projection.

The generating function $\mathrm{GF}(G, \{\text{cycles}\})$ of an edge weighted graph G equals the sum of the weights of all Hamilton cycles in G. In particular, $\mathrm{UHC}_n = \mathrm{GF}(K_n, \{\text{cycles}\})$.

Theorem 3.18 *There is a p-sequence of planar, edge weighted graphs G_n with maximal node degree 3 such that $(\mathrm{GF}(G, \{\text{cycles}\}))$ is complete.*

This result has the following consequence.

Corollary 3.19 *The sequences (UHC_n) and (HC_n) of undirected and directed Hamilton cycle polynomials, respectively, are both complete.*

The claim about (UHC_n) follows immediately from Thm. 3.18. In order to show the claim for (HC_n), replace each edge $e = \{i, j\}$ of K_n by the two directed edges (i, j) and (j, i) of weight X_e.

For the proof of Thm. 3.18 we need two auxiliary constructions. Consider the graph C in Fig. 3.3 with the connector nodes a, b and the distinguished edges e and f. One edge indicated in the figure has the weight $1/2$ and all other edges are supposed to be of weight 1. The following facts about Hamilton paths in C between a and b are easily verified. There are exactly two of them

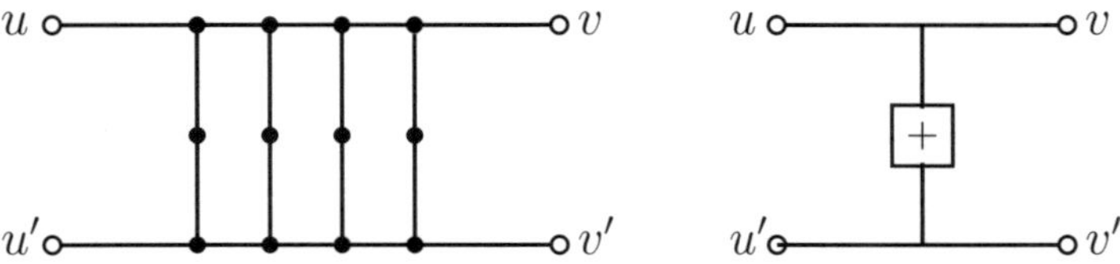

Figure 3.4: The exclusive-or coupling of edges and its symbol.

which avoid both edges e and f, and each of them has the weight $1/2$. There is exactly one Hamilton path which takes edge e but avoids f; it has the weight 1. The same is true for the Hamilton path taking f but avoiding e. Finally, there is no Hamilton path which uses both edges e and f. To summarize, the sum of the weights of all Hamilton paths with prescribed behaviour on the edges e and f equals 0 if both of these edges have to be taken, and equals 1 otherwise.

We also need the well-known *exclusive-or coupling* in Fig. 3.4 from [39]. Built in between two edges $e = \{u, v\}$ and $e' = \{u', v'\}$ of a graph, it has the effect to force that a Hamilton cycle has to take exactly one of the edges e or e'. Schematically, the coupling is represented by the symbol in Fig. 3.4. We suppose that all edges occuring in it carry the weight 1.

For the planarity it is crucial that two crossing exclusive-or couplings can be replaced by a planar construction sharing the same properties. This is sketched in Fig. 3.5.

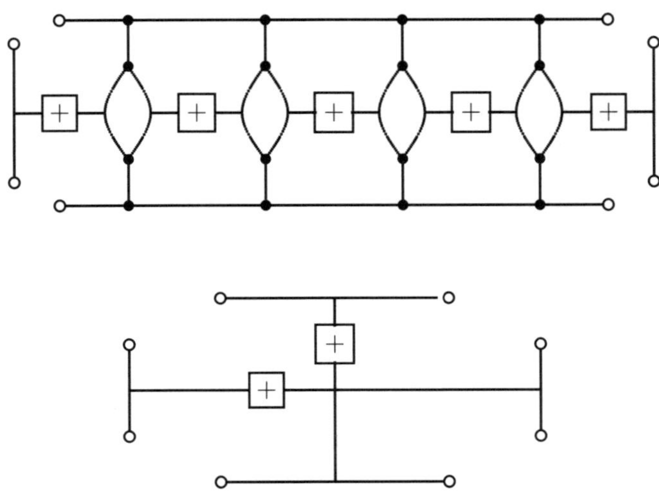

Figure 3.5: A planar construction for two crossing exclusive-or couplings.

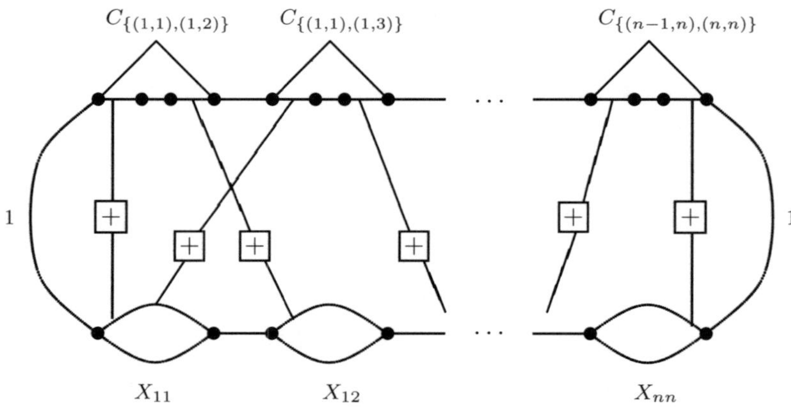

Figure 3.6: The graph G_n.

Proof. (of Thm. 3.18) Consider the complete bipartite graph $K_{n,n}$. For each edge $e = \{i, j\}$ of $K_{n,n}$ we create the following graph V_e

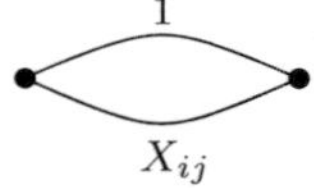

with edge weights as indicated. The edge with weight 1 will be called the *upper edge* and the other the *lower edge* of V_e. Moreover, for each pair of edges $\{e, f\}$ sharing a node, we create a copy $C_{\{e,f\}}$ of our auxiliary graph C and call its distinguished edges also e and f. Note that we create $2n\binom{n}{2}$ copies of C this way. We connect all these graphs $C_{\{e,f\}}$ and V_e at their connector nodes to form a cycle as in Fig. 3.6. Finally, for each copy $C_{\{e,f\}}$ we introduce an exclusive-or coupling between the edge e of $C_{\{e,f\}}$ and the upper edge of V_e, as well as an exclusive-or coupling between the edge f of $C_{\{e,f\}}$ and the upper edge of V_f. For crossing exclusive-or couplings we insert the construction of Fig. 3.5. The resulting edge weighted graph G_n is planar by construction. It is easy to check that each node of G_n has degree at most 3 and that the number of nodes of G_n is p-bounded in n.

By Thm. 3.7 it is now sufficient to verify the following claim

$$\mathrm{PER}_n^* = \mathrm{GF}(G_n, \{\text{cycles}\}) \ .$$

Suppose σ is a Hamilton cycle of the graph G_n. We assign to σ the set $\pi = \pi(\sigma)$ of those edges e of $K_{n,n}$, where σ chooses the lower edge of V_e. Then π must be a partial matching of $K_{n,n}$. To see this, assume that there are distinct edges $e, f \in \pi$ of $K_{n,n}$ sharing a node. By the exclusive-or coupling, the Hamilton cycle σ has to take both edges e and f of the graph $C_{\{e,f\}}$,which is impossible.

On the other hand, let π be any partial matching of $K_{n,n}$. Suppose that σ is a Hamilton cycle of G_n satisfying $\pi(\sigma) = \pi$. By the exclusive-or coupling, it is prescribed for each of the graphs $C_{\{e,f\}}$ which of its edges e or f have to be taken. However, it is never required that both have to be taken. By the properties of the graph C we therefore obtain that the sum of the weights of all Hamilton cycles σ satisfying $\pi(\sigma) = \pi$ equals $\prod_{e \in \pi} X_e$.

From these considerations we conclude that

$$\mathrm{GF}(G_n, \{\text{cycles}\}) = \sum_\sigma w(\sigma) = \sum_\pi \sum_{\pi(\sigma)=\pi} w(\sigma) = \sum_\pi \prod_{e \in \pi} X_e = \mathrm{PER}_n^* \ ,$$

where σ runs over all Hamilton cycles of G_n and π runs over all partial matchings of $K_{n,n}$. $\qquad\square$

Remark 3.20 By some modifications, we can additionally achieve that the graphs G_n in Thm. 3.18 are bipartite. The graph C becomes bipartite if we insert a node on the middle edge joining e and f. The exclusive-or couplings and their crossings can be realized as bipartite, planar graphs by adding "horizontal rows" of four nodes, if necessary. We leave the details to the interested reader.

3.3.6   Self Avoiding Walks

In Sect. 3.1 we assigned to an edge weighted graph G and a graph property $\mathcal{E}$ a generating function $\mathrm{GF}(G, \mathcal{E})$. We can do the same for graphs G with two distinguished nodes s, t and properties of such objects. Consider the property that a graph is a simple path between two distinguished nodes s and t. We call such a path an s-t-path or s-t-self avoiding walk in the sequel. The generating function corresponding to this property enumerates the Hamilton paths between s and t in a given graph G. We will denote it by $\mathrm{HP}(G)$. If we focus on the set of graphs consisting of an s-t-path and further isolated nodes, then the corresponding generating function $\mathrm{SW}(G)$ enumerates the s-t-paths in the graph G.

Valiant [109] has shown that counting the number of s-t-paths in a given graph is a #P-complete problem. We prove here a corresponding algebraic result for planar graphs.

Theorem 3.21 *Let k be infinite. There is a p-sequence of planar, edge weighted graphs G_n with maximal node degree 3 such that $(\mathrm{SW}(G_n))$ is complete.*

Before proving this theorem, we proceed with some general considerations. By using interpolation, one sometimes establishes that a polynomial f_n is a projection of a linear combination of polynomially many polynomials g_m. In general, it is not clear whether this linear combination is itself a projection of some g_M, where M is p-bounded in n. We capture this observation by the following definition.

Definition 3.22 We call a p-family $f = (f_n)$ a *linear p-projection* of a p-family $g = (g_n)$, in symbols $f \leq_p^\ell g$, iff there is some p-bounded $t\colon \mathbb{N} \to \mathbb{N}$ such that for all n there are $i_1 \leq i_2 \leq \ldots \leq i_{t(n)} \leq t(n)$ and $\lambda_\tau \in k$ such that f_n is a projection of the linear combination $\sum_{\tau=1}^{t(n)} \lambda_\tau g_{i_\tau}$, where the sets of variables of the g_{i_τ} are thought to be (made) disjoint for distinct τ.

It is easy to check that the linear p-projection is transitive. If $\leq_c$ denotes the c-reduction to be introduced in Sect. 5.4, we have

$$f \leq_p g \implies f \leq_p^\ell g \implies f \leq_c g \ .$$

The notions of VNP-completeness with respect to p-projection and linear p-projection coincide for certain families.

Definition 3.23 A p-family (g_n) is called *linearly closed* iff any linear combination $\sum_{\tau=1}^n \lambda_\tau g_{i_\tau}$ is a projection of some g_m, where m is p-bounded in the number n of terms and $\max_\tau i_\tau$. Hereby, the sets of variables of the g_{i_τ} are supposed to be (made) disjoint for distinct τ.

Proposition 3.24 *Let g be p-definable. Then g is VNP-complete (with respect to p-projection) if and only if g is VNP-complete with respect to linear p-projection, and linearly closed.*

In particular, each of the families which we have proven to be complete, is linearly closed. Sometimes, it is easy to directly verify the linear closedness property. We invite the reader to to this for the clique family. For other examples, for instance the permanent family, this is a much more challenging exercise.

Problem 3.2 Is the determinant family linearly closed?

Instead of directly proving completeness with respect to p-projection, it is sometimes more convenient to show completeness with respect to linear p-projection, and linear closedness separately. For the family $(\mathrm{SW}(K_n))$, the linear closedness is almost obvious. Namely, let G_i be an edge weighted graph with distinguished nodes s_i, t_i of degree 1 and suppose $\lambda_i \in k$ $(i = 1, \ldots, n)$. We form the disjoint union of $G_1, \ldots, G_n$ and additional nodes $s_1', \ldots, s_n', s, t$ and introduce edges $\{s_i', s_i\}$ of weight λ_i. Moreover, we connect each of the sequences of nodes $s_1', \ldots, s_n', s$ and $t_1, \ldots, t_n, t$ by a line (consisting of edges of weight 1). It is clear that the resulting graph G satisfies

$$\mathrm{SW}(G) = \sum_{i=1}^n \lambda_i \mathrm{SW}(G_i) \ .$$

This already shows that the family $(\mathrm{SW}(K_n))$ is linearly closed. We remark that if all G_i are planar with maximal node degree 3, then G is also planar with maximal node degree 3.

Proof. (of Thm. 3.21) We use the interpolation technique in Valiant [109].

From the proof of Thm. 3.18 we know that there is a p-sequence of planar, edge weighted graphs G_n with maximal node degree 3 and distinguished nodes s, t of degree 1 such that $(\mathrm{HP}(G_n))$ is complete.

We modify G_n by subdividing each of its edges e by an additional node into two edges. One of them is supposed to carry the weight $w(e)$, and the other the indeterminate weight X. Let us call the resulting graph $G'_n(X)$. Apparently, the s-t-paths π in G_n are in bijective correspondence with the s-t-paths π' in $G'_n(X)$. Moreover, $w(\pi') = X^\ell w(\pi)$ if π has length ℓ. If SW_ℓ denotes the generating function for s-t-paths of length ℓ, we have

$$\mathrm{SW}(G'_n(X)) = \sum_{\ell=1}^{N_n-1} X^\ell \, \mathrm{SW}_\ell(G_n) \;,$$

where N_n is the number of nodes of G_n. We may choose distinct elements $\alpha_1, \ldots, \alpha_{N_n-1} \in k$ as k is supposed to be infinite. The Vandermonde matrix $[\alpha_\nu^\ell]$ is invertible, hence there exist $\beta_\nu \in k$ such that

$$\mathrm{HP}(G_n) = \mathrm{SW}_{N_n-1}(G_n) = \sum_{\nu=1}^{N_n-1} \beta_\nu \, \mathrm{SW}(G'_n(\alpha_\nu)) \;.$$

We conclude that $(\mathrm{HP}(G_n))$ is a linear p-projection of the family $(\mathrm{SW}(G'_n(X)))$. By the reasonings from before on linear closedness, the theorem is proved. $\square$

Problem 3.3 Is (SW_n) complete over finite fields?

3.3.7 Connectivity

We recall the notion of the probability generating function

$$\mathrm{PGF}(G, \mathcal{E}) := \sum_{E' \subseteq E} P(E')$$

corresponding to an edge weighted graph $G = (V, E)$ and a graph property $\mathcal{E}$, which was introduced in Sect. 3.1. The quantity

$$P(E') := \prod_{e \in E'} w(e) \prod_{e \in E \setminus E'} (1 - w(e))$$

can be interpreted as the probability of the elementary event E', if the edges e in E are picked independently at random with probability $w(e)$. Thus $\mathrm{PGF}(G, \mathcal{E})$ is the probability that the resulting random subgraph of G has the property $\mathcal{E}$.

We will study the following two graph properties: the property $\mathcal{CO}$ of a graph to be connected, and the property $\mathcal{STC}$ that two distinguished nodes s and t in a graph can be connected by an s-t-path.

Jerrum [52] has shown that the probability generating functions of the properties $\mathcal{CO}$ and $\mathcal{STC}$ corresponding to complete graphs yield complete families. Let us call these generating functions *connectivity polynomials*.

Theorem 3.25 (Jerrum) *The families* $(\mathrm{PGF}(K_n, \mathcal{CO}))$ *and* $(\mathrm{PGF}(K_n, \mathcal{STC}))$ *of connectivity polynomials are complete over* $\mathbb{Q}$.

We remark that Valiant [109] had proven before that computing the probability that a random graph contains an s-t-path is #P-hard. For related results see also Provan and Ball [92].

Provan [91] has shown that computing the probability that a random graph contains an s-t-path remains #P-hard for *planar* graphs. In a similar way, we will prove the following result.

Theorem 3.26 *There is a p-sequence of planar graphs G_n with distinguished nodes s, t and maximal node degree 3 such that $(\mathrm{PGF}(G_n, \mathcal{STC}))$ is complete over* $\mathbb{Q}$.

We will demonstrate this theorem by showing completeness with respect to linear p-projection and linear closedness separately. Let us first verify the linear closedness property.

Lemma 3.27 *Let G_i be an edge weighted graph with distinguished nodes s_i, t_i of degree 1 and suppose $\lambda_i \in k$ for $1 \leq i \leq n$. Then there is an edge weighted graph G with distinguished nodes s, t, whose number of nodes is p-bounded in n and the number of nodes of all G_i, such that*

$$\mathrm{PGF}(G, \mathcal{STC}) = \sum_{i=1}^{n} \lambda_i \mathrm{PGF}(G_i, \mathcal{STC}) \ .$$

In particular, the family $(\mathrm{PGF}(K_n, \mathcal{STC}))$ is linearly closed. Moreover, if all G_i are planar with maximal node degree 3, then G may be chosed to be planar with maximal node degree 3 as well.

Proof. We first treat the case $n = 2$. Consider the edge weighted graph $A = (U, F)$ of Fig. 3.7. Assume the edges of A are independently chosen with

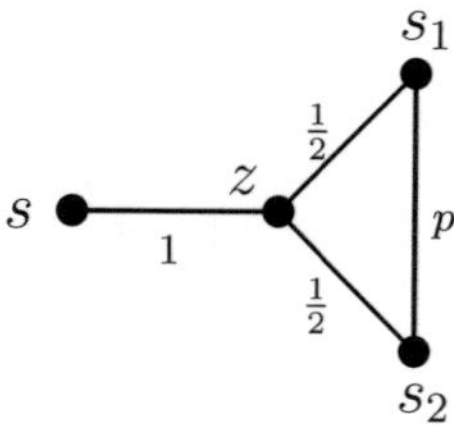

Figure 3.7: The auxiliary graph $A = (U, F)$.

the indicated probabilities. Then we have

$$
\begin{aligned}
P_1 \;&=\; \text{Prob}\{s, s_1 \text{ in same connected component, } s_2 \text{ in different component}\}\\
&=\; (1-p)/4 \;,\\
P_2 \;&=\; \text{Prob}\{s, s_1, s_2 \text{ in same connected component}\}\\
&=\; p/4 + p/4 + (1-p)/4 + p/4 = (1+2p)/4 \;.
\end{aligned}
$$

By choosing $p = -1/2$ we achieve $P_1 = 3/8$ and $P_2 = 0$. (Although this choice of p is not a realizable probability, it is an admissible substitution.) Our intention is to use the graph A as a device which forces that either s and s_1, or s and s_2 are in the same connected component.

We construct now an edge weighted graph G from $G_i = (V_i, E_i)$, $i = 1, 2$, and $A = (U, F)$ according to Fig. 3.8. Let $H := \{\{t, t_1\}, \{t, t_2\}\}$. For a subset $F' \subseteq F$ we set $P(F') = \prod_{e \in F'} w(e) \prod_{e \in F \setminus F'} (1 - w(e))$, and we define the quantities $P(E_i')$ and $P(H')$ for subsets $E_i' \subseteq E_i$ and $H' \subseteq H$ similarly. It is easy to check that

$$
\mathrm{PGF}(G, \mathcal{STC}) = \sum P(F')P(E_1')P(E_2')P(H') \;,
$$

where the sum is over all subsets $F' \subseteq F$, $E_i' \subseteq E_i$, $H' \subseteq H$ such that the set of edges $F' \cup E_1' \cup E_2' \cup H'$ contains an s-t-path. The above sum splits as $\mathrm{PGF}(G, \mathcal{STC}) = I_1 + I_2,$ where

$$
I_1 = P\{\{s,z\}, \{z,s_1\}\}\Big(\textstyle\sum_{E_1'} P(E_1')\Big)\Big(\textstyle\sum_{E_2'} P(E_2')\Big)\Big(\textstyle\sum_{H'} P(H')\Big) \;.
$$

Here, the first sum is over all E_1' such that (V_1, E_1') contains an s_1-t_1-path, hence it equals $\mathrm{PGF}(G_1, \mathcal{STC})$. The second sum is over the subsets $E_2' \subseteq E_2$ without restriction and equals therefore $\prod_{e \in E_2}(w(e) + (1 - w(e))) = 1$. The third sum equals $8\lambda_1/3$. I_2 is defined analogously. (Recall that the "probability" P_2 that s, s_1, s_2 are in the same connected component is zero by construction of A.)

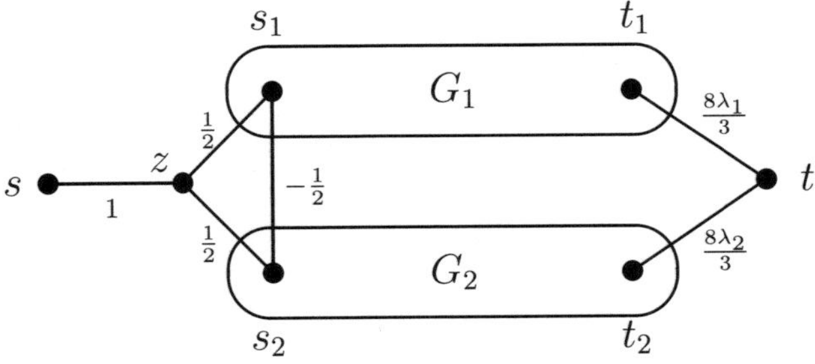

Figure 3.8: The graph G.

Therefore, we have $I_1 = \lambda_1 \text{PGF}(G_1, \mathcal{STC})$ and similarly we get that $I_2 = \lambda_2 \text{PGF}(G_2, \mathcal{STC})$. We conclude that

$$\text{PGF}(G, \mathcal{STC}) = \lambda_1 \text{PGF}(G_1, \mathcal{STC}) + \lambda_2 \text{PGF}(G_2, \mathcal{STC}) \ .$$

If b is an upper bound on the number of nodes of G_i, then G has at most $2b + 3$ nodes.

To settle the general case, we may assume that $n = 2^s$. A recursive application of the construction before yields a desired graph G_s on n_s nodes satisfying $n_{s+1} \le 2n_s + 3$. Therefore, $n_{s+1} \le 2^s b + 3(2^s - 1) \le n(b+3)$, which is polynomial in n and b. $\qquad\square$

For proving the completeness with respect to linear p-projection we need the following formula.

Lemma 3.28 *Let $G = (V, E)$ be an edge weighted graph with distinguished nodes s, t. Then we have*

$$\text{PGF}(G, \mathcal{STC}) = \sum_H \epsilon_H w(H) \ ,$$

where the sum is over all sets of edges H belonging to a nonempty collection of s-t-paths, and ϵ_H are integers depending only on H. Moreover, $\epsilon_H = 1$ if H is the set of edges of a single s-t-path.

One can explicitly describe the ϵ_H, which turn out to be either -1, 0, or 1 (cf. Satyanarayana and Prabhakar [94] or Willie [119]), but we will not need this in the sequel.

Proof. It suffices to show the assertion in the situation where the weights $w(e)$ are independent indeterminates. To verify the claimed polynomial identity, we may then assume that $w(e) \in [0, 1]$ and interpret this as the probability that the edge $e \in E$ is chosen.

Let $\pi_1, \ldots, \pi_t$ be the edge sets of all s-t-paths in G, and consider the events $A_i := \{E' \subseteq E \mid \pi_i \subseteq E'\}$. Apparently, $\text{PGF}(G, \mathcal{STC})$ equals the probability of the event $A_1 \cup \ldots \cup A_t$. By the principle of inclusion and exclusion we may write

$$\text{Prob}(A_1 \cup \ldots \cup A_t) = \sum_{j_1 < \ldots < j_t} (-1)^{t-1} \text{Prob}(A_{j_1} \cap \ldots \cap A_{j_t}) \ .$$

Note that for a union $H = \pi_{j_1} \cup \ldots \cup \pi_{j_t}$ of s-t-paths we have

$$\text{Prob}(A_{j_1} \cap \ldots \cap A_{j_t}) = \text{Prob}\{E' \mid H \subseteq E'\} = \prod_{e \in H} w(e) = w(H) \ .$$

By collecting the terms yielding the same edge set H, we get the following sum over all unions H of s-t-paths

$$\text{Prob}(A_1 \cup \ldots \cup A_t) = \sum_H (\epsilon_H^+ - \epsilon_H^-) w(H) \ ,$$

where ϵ_H^+ equals the number of tuples $j_1 < \ldots < j_t$ of odd length t such that $H = \pi_{j_1} \cup \ldots \cup \pi_{j_t}$, and ϵ_H^- is the analogous number of tuples of even length t. If $H = \pi_{j_1}$ is a single s-t-path, then it is clear that $\epsilon_H^+ = 1$ and $\epsilon_H^- = 0$. This proves the assertion. $\qquad\square$

Proof. (of Thm. 3.26) From the proof of Thm. 3.18 we know that there is a p-sequence of planar, edge weighted graphs G_n with maximal node degree 3 and two distinguished nodes s, t of degree 1 such that $(\mathrm{HP}(G_n))$ is complete.

Let g_{ni} denote the number of nodes of G_n of degree i. We have $g_{ni} = 0$ for $i > 3$. Moreover, we may assume w.l.o.g. that $g_{n1} = 2$, since otherwise $\mathrm{HP}(G_n) = 0$.

We modify now the graph G_n as follows. Let X, Y be new indeterminates. In a first step, we subdivide each edge e of G_n by an additional node into two edges. One of them is supposed to carry the weight $w(e)$, and the other the indeterminate weight X. In a second step, we replace each node of G_n of degree 3 by a triangle as follows:

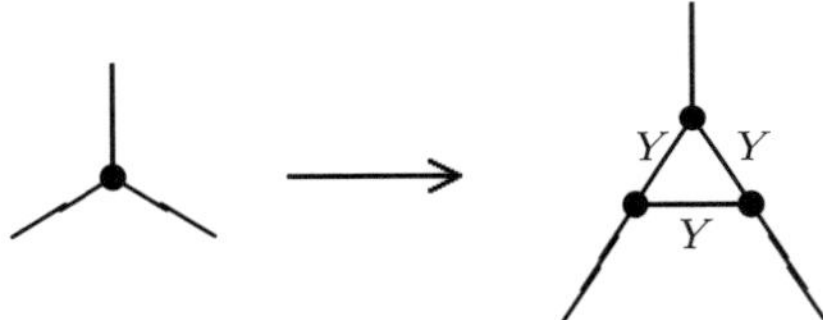

The edges of the triangle are supposed to carry the indeterminate weight Y. We denote the resulting planar, edge weighted graph by $G_n' = G_n'(X, Y)$ and write

$$P_n(X, Y) := \sum_{i,j} Q_{ij}^n X^i Y^j := \mathrm{PGF}(G_n', \mathcal{STC}) \ .$$

To any set of edges H' belonging to a collection of s-t-paths of G_n', there corresponds a set of edges H of G_n in a natural way. H obviously belongs to a collection of s-t-paths of G_n. We introduce the following notations:

$$
\begin{aligned}
i(H') \quad &= \quad \#\{\text{edges of } H' \text{ of weight } X\} = \#\{\text{edges of } H\} \ , \\
j(H') \quad &= \quad \#\{\text{edges of } H' \text{ of weight } Y\} = \#\{\text{triangle edges of } H'\} \ .
\end{aligned}
$$

Note that $w(H') = w(H) X^{i(H')} Y^{j(H')}$. By Lemma 3.28 we have

$$Q_{ij}^n = \sum_{H'} \epsilon_{H'} \, w(H) \ ,$$

where the sum is over all sets of edges H' belonging to a collection of s-t-paths such that $i = i(H')$ and $j = j(H')$. The degree of $P_n(X, Y)$ in X is bounded by the number γ_n of edges of G_n, and the degree in Y is most $3g_{n3}$. Choose distinct rational numbers $\xi_0, \ldots, \xi_{\gamma_n}$ and $\eta_0, \ldots, \eta_{3g_{n3}}$. It is clear that the coefficients Q_{ij}^n of $P_n(X, Y)$ can be written as a linear combination of the values $P_n(\xi_\sigma, \eta_\rho)$. Thus the sequence of all Q_{ij}^n is a linear p-projection of $(P_n(X, Y))$.

We now focus on the coefficients Q_{i_n, j_n}^n corresponding to the degrees

$$i_n := |G'_n| - 1 = 1 + g_{n2} + g_{n3}, \quad j_n := g_{n3} \ .$$

Note that any s-t-Hamilton path H of G_n defines an s-t-path H' of G'_n satisfying $i(H') = i_n$ and $j(H') = j_n$. (In each triangle exactly one edge is taken.) We will prove the following crucial

Claim: Any edge set H' belonging to a collection of s-t-paths, which satisfies $i(H') = i_n$ and $j(H') = j_n$, must necessarily be an s-t-path of the above form.

This claim implies that we have a bijection between such H' and the s-t-Hamilton paths H of G_n. Therefore,

$$Q_{i_n, j_n}^n = \sum_H w(H) = \mathrm{HP}(G_n) \ ,$$

as $\epsilon_{H'} = 1$ for s-t-paths H'. We conclude that $(\mathrm{HP}(G_n))$ is a linear p-projection of $(P_n(X, Y))$, hence the latter family is VNP-complete with respect to linear p-projections. Lemma 3.27 implies now the assertion of the theorem (compare Prop. 3.24).

To prove the claim, consider an edge set H' belonging to a collection of s-t-paths of G'_n such that $i(H') = i_n$ and $j(H') = j_n$. Consider the spanning subgraph $G(H)$ of G_n induced by the edge set H corresponding to H'. Let h_i denote the number of nodes of degree i of $G(H)$. It is easy to see that $h_1 = 2, h_2 \leq g_2, h_3 \leq g_3$, and $h_i = 0$ for $i > 3$. Moreover, we have

$$2 + 2h_2 + 3h_3 = 2\,i(H') = 2i_n = 2 + 2g_{n2} + 2g_{n3} \ . \qquad .$$

For each node of $G(H)$ of degree 3 at least two edges of the corresponding triangle are contained in H', as H' is a union of s-t-paths. For each node of $G(H)$ of degree 2, which has degree 3 in G_n, at least one edge of the corresponding triangle is contained in H'. This implies the inequality

$$(h_2 - g_{n2}) + 2h_3 \leq j(H') = g_{n3} \ .$$

We conclude that $2h_2 + 3h_3 \geq 2g_{n2} + 2(h_2 - g_{n2}) + 4h_3 = 2h_2 + 4h_3$, hence $h_3 = 0$. Therefore, H must be an s-t-path. As its number of edges equals i_n, it is in fact an s-t-Hamilton path. Hence H' is an s-t-path in G'_n which takes in each triangle exactly one edge. This proves the claim. $\qquad\square$

4

Cook's versus Valiant's Hypothesis

We first investigate the dependency of our complexity concepts on the field k. We prove that for algebraically closed fields k, the truth of Valiant's hypothesis $\mathrm{VP} \neq \mathrm{VNP}$ depends at most on the characteristic of k.

We then relate the hypothesis $\mathrm{VP} \neq \mathrm{VNP}$ to hypotheses in classical discrete complexity theory. We identify the Boolean parts of VP and VNP as familiar nonuniform discrete complexity classes. As a consequence, we obtain rather strong evidence for Valiant's hypothesis: if it were wrong, then the nonuniform versions of the parallel complexity class NC and the polynomial hierarchy PH would be equal! In particular, the polynomial hierarchy would collapse to the second level. We prove this for fields of characteristic zero and finite fields; in the first case we assume the generalized Riemann hypothesis. The main results of this chapter are taken from Bürgisser [19].

4.1　Dependence on the Field

Let (f_n) be a p-family over k. Is it posssible that (f_n) is p-computable over an extension field K of k, but not p-computable over k? We will see that this phenomenon cannot happen for certain field extensions $k \subseteq K$.

Let $L_k(f)$ denote the complexity to compute f from variables and constants in k (no divisions). It is clear that $L_K(f) \leq L_k(f)$ if K is an extension field of k.

Part (i) and (ii) of the following proposition go back to Strassen [103] (see also [21, Thm. 4.17]).

Proposition 4.1 *Let $f \in k[X_1, \ldots, X_m]$ and $k \subseteq K$ be a field extension. Then we have:*

(i) $L_k(f) = L_K(f)$ if k is algebraically closed.

(ii) $L_k(f) = L_K(f)$ if $k \subseteq K$ is a purely transcendental extension of the infinite field k.

(iii) $L_k(f) = O(d^3 L_K(f))$ if $k \subseteq K$ is a finite algebraic extension of degree d. If the extension is separable, then d^3 can be replaced by the complexity $M(d)$ for multiplying two polynomials of degree d over k symbolically.

Proof. (i) Let $a_1, \ldots, a_m \in K$ be the constants used by an optimal straight-line program computing f over K. By Hilbert's Nullstellensatz (cf. Lang [73, p. 375]), there is a k-algebra morphism $\varphi \colon A = k[a_1, \ldots, a_m] \to k$, since k is assumed to be algebraically closed. The same straight-line program computes $f = \varphi(f)$ from the variables and the constants $\varphi(a_i)$. Thus $L_k(f) = L_K(f)$.

(ii) An optimal computation of f over $K = k(Y_1, \ldots, Y_m)$ actually takes place in a localization $A = \{g/N^s \mid g \in k[Y_1, \ldots, Y_m], s \in \mathbb{N}\}$, where N is a nonzero polynomial in the Y_i. As k is infinite, there is some $\eta \in k^m$ such that $N(\eta) \neq 0$. Now apply the k-algebra morphism

$$A \to k, \; g/N^s \mapsto g(\eta)/N(\eta)^s$$

and finish the argumentation as before.

(iii) We may represent the elements of K as vectors of length d over k. An addition or multiplication of elements of K may be simulated by $O(d^3)$ arithmetic operations of elements of k. If the extension is separable, then we have a simple field extension $K \simeq k[T]/(g)$, and arithmetic operations in K may be simulated with $O(M(d))$ arithmetic operations over k (cf. [21, Sect. 2]). $\qquad\qquad\square$

As a consequence, we obtain the following result about the dependency of our complexity concepts on the field. (For the definition of c-reductions see Sect. 5.4.)

Corollary 4.2 *Assume we are in one of the following cases: (i) k algebraically closed, (ii) $k \subseteq K$ purely transcendental field extension and k infinite, (iii) $k \subseteq K$ finite algebraic field extension. Then we have for all p-families (f_n) over k:*

(1) *(f_n) p-computable over K $\Rightarrow$ (f_n) p-computable over k.*

(2) *(f_n) p-definable over K $\Rightarrow$ (f_n) p-definable over k.*

(3) *If Valiant's hypothesis is true over k, then it is true over K.*

(4) *If Valiant's hypothesis is true over a field, then is is true over all of its subfields.*

(5) *If (f_n) is a p-family over k, which is VNP_K-complete with respect to p-projections, then it is also VNP_k-complete with respect to p-projections in the above cases (i) and (ii). In the case (iii), we conclude only completeness with respect to c-reductions.*

Proof. Claim (1) is clear by Prop. 4.1.

(2) Let (g_n) be p-computable over K and assume $f_n(X) = \sum_e g_n(X, e)$. In the cases (i) and (ii), we may argue as in the proof of Prop. 4.1 and find for each n some k-algebra morphism $\varphi_n \colon A_n \to k$ satisfying $L_k(\varphi(g_n)) \leq L_K(g_n)$.

Then $(\varphi(g_n))$ is p-computable over k and $f_n(X) = \sum_e \varphi(g_n)(X, e)$. Thus (f_n) is p-definable over k.

In the case (iii) assume that $1 = \kappa_1, \ldots, \kappa_d$ is a basis of the k-vector space K. Write $g_n = \sum_{\delta=1}^d g_{n,\delta}\kappa_\delta$, where $g_{n,\delta}$ are polynomials over k. Arguing as in the proof of Prop. 4.1, we may conclude that $L_k(g_{n,1}) = O(d^3 L_K(g_n))$. Thus $(g_{n,1})$ is p-computable over k. Moreover, we have $f_n(X) = \sum_e g_{n,1}(X, e)$, hence (f_n) is p-definable over k.

(3) follows immediately from (1).

(4) Let $E \subseteq F$ and assume we had $\mathrm{VP}_E = \mathrm{VNP}_E$. Then $L_E(\mathrm{HC}_n)$ would be p-bounded. As trivially $L_F(\mathrm{HC}_n) \leq L_E(\mathrm{HC}_n)$, the family $L_F(\mathrm{HC}_n)$ would be p-bounded as well. Since the Hamilton cycle family (HC_n) is VNP_F-complete, we would have $\mathrm{VP}_F = \mathrm{VNP}_F$.

(5) Suppose (f_n) is VNP-complete over K. Let (g_n) be any p-definable family over k. Then there is some $t: \mathbb{N} \to \mathbb{N}$, p-bounded from above and below, such that g_n is a projection of $f_{t(n)}$ over K, that is,

$$g_n(X) = f_{t(n)}(a^{(n)}) \ ,$$

where the components of $a^{(n)}$ are variables X_i or constants in K. Let $A_n \subseteq K$ be the k-algebra generated by the constants occuring in $a^{(n)}$. We assume we are in the case (i) or (ii). As in the proof of Prop. 4.1, we see that there is a k-algebra morphism $\varphi_n: A_n \to k$. If we denote its extension $A_n[X] \to k[X]$ also by φ_n, we have $g_n(X) = f_{t(n)}(\varphi_n(a^{(n)}))$, thus g_n is a projection of $f_{t(n)}$ over k. Hence (f_n) is VNP-complete over k. We leave the easy, but somewhat tedious argumentation in the case (iii) to the reader. $\square$

We formulate now the following *transfer theorem*, which is the main insight of this section. Let $\overline{\mathbb{F}}_p$ denote the algebraic closure of the prime field $\mathbb{F}_p$ of characteristic $p > 0$ and put $\mathbb{F}_0 := \mathbb{Q}$.

Theorem 4.3 *Let $p \geq 0$. The following statements are equivalent:*

(1) *Valiant's hypothesis is true over all fields of characteristic p.*

(2) *Valiant's hypothesis is true over some algebraically closed field of characteristic p.*

(3) *Valiant's hypothesis is true over $\overline{\mathbb{F}}_p$.*

Proof. $(1) \Rightarrow (2)$ is trivial.

$(2) \Rightarrow (3)$ follows from Cor. 4.2(4), as $\overline{\mathbb{F}}_p$ is contained in any algebraically closed field of characteristic p.

$(3) \Rightarrow (1)$: Let $\mathbb{F}_p \subseteq k$ be a field extension. By Cor. 4.2(3), Valiant's hypothesis is true over the algebraic closure $\overline{k}$ of k. Thus is also true over the subfield k of $\overline{k}$. $\square$

Problem 4.1 Does Valiant's hypothesis over the rationals $\mathbb{Q}$ necessarily imply Valiant's hypothesis over the field of algebraic numbers $\overline{\mathbb{Q}}$?

4.2   Statement of Main Results

Our goal is to establish close relations between Valiant's algebraic model and discrete complexity theory, based on the computational model of Turing machines. In particular, we will show that Valiant's hypothesis is a consequence of standard hypotheses in discrete complexity theory.

The *weight* $\mathrm{wt}(f)$ of a polynomial $f \in \mathbb{Z}[X_1, \ldots, X_n]$ is defined as the sum of the absolute values of its coefficients. Let $f_1, \ldots, f_s \in \mathbb{Z}[X_1, \ldots, X_n]$ be polynomials of degree and weight bounded by $d \geq n$ and w, respectively. We assume that the system of equations

$$(S) \qquad\qquad f_1 = 0, \ f_2 = 0, \ldots, f_s = 0$$

is solvable over $\mathbb{C}$. Let $\pi(x)$ denote the number of primes $\leq x$, and let $\pi_S(x)$ stand for the number of primes $p \leq x$ such that (S) is solvable over the finite field $\mathbb{F}_p$.

Our main tool is Thm. 4.4 below, which is of considerable interest in its own right. It is an improvement of Thm. 8 in Koiran [67], based on the results in Krick and Pardo [68, 69]. (GRH) denotes the generalized Riemann hypothesis for number fields.

Theorem 4.4 *Assuming the generalized Riemann hypothesis (GRH), we have*

$$\pi_S(x) \geq \frac{\pi(x)}{d^{O(n)}} - x^{1/2} \log(wx)$$

for all systems (S) solvable over $\mathbb{C}$.

We are going to assign to the complexity classes VP_k and VNP_k their "Boolean parts" which consist of certain string functions $\{0,1\}^* \to \{0,1\}^*$. Let $f = (f_n)$ be a p-family over k such that $f_n \in k[X_1, \ldots, X_n]$. If $\mathrm{char} k = 0$, we assume that $f_n(x)$ is a natural number of bitsize $n^{O(1)}$ for all $x \in \{0,1\}^n$. If $\mathrm{char} k = p > 0$, we assume that the coefficients of f_n are contained in the prime field $\mathbb{F}_p$ of k. We define the Boolean part of such an f as the string function which maps $x \in \{0,1\}^n$ to the binary encoding of $f_n(x)$. The Boolean part $\mathrm{BP}(\mathrm{VP}_k)$ of VP_k is the set of the Boolean parts of all $f \in \mathrm{VP}_k$ for which it is defined. The Boolean part $\mathrm{BP}(\mathrm{VNP}_k)$ is defined analogously.

Our main result identifies the Boolean parts of VP_k and VNP_k as familiar nonuniform discrete complexity classes. For a review of these classes, see Sect. 4.3.

Theorem 4.5 *(A) Under (GRH) we have for fields k of characteristic zero*

$$\begin{array}{ccccc}
\mathrm{FNC}^1/\mathrm{poly} & \subseteq & \mathrm{BP}(\mathrm{VP}_k) & \subseteq & \mathrm{FNC}^3/\mathrm{poly}, \\
\#\mathrm{P}/\mathrm{poly} & \subseteq & \mathrm{BP}(\mathrm{VNP}_k) & \subseteq & \mathrm{FP}^{\#\mathrm{P}}/\mathrm{poly} \ .
\end{array}$$

(B) *For finite fields of characteristic p we have*

$$\mathrm{FNC}^1/\mathrm{poly} \subseteq \mathrm{BP}(\mathrm{VP}_k) \subseteq \mathrm{FNC}^2/\mathrm{poly}, \quad \#_p\mathrm{P}/\mathrm{poly} = \mathrm{BP}(\mathrm{VNP}_k) \ .$$

We conjecture that statement (A) can be strengthened to $\#\mathrm{P}/\mathrm{poly} = \mathrm{BP}(\mathrm{VNP}_k)$.

The main difficulty in proving this theorem comes from the fact that Valiant's model allows the use of arbitrary constants in k. Assume we had a straight-line program Γ_n of polynomial size, which computes the permanent $\mathrm{per}(A)$ of a given matrix A using some complex constants. In order to transform Γ_n into a polynomial size Boolean circuit which computes the permanent of $0,1$-matrices, we would have to replace the complex constants by something discrete of small size. In fact, we can show that for each n there is a large prime number $p_n > n!$ of polynomial bitsize, and that there are constants in $\mathbb{F}_{p_n}$, such that the simulation of Γ_n on these constants in $\mathbb{F}_{p_n}$ computes $\mathrm{per}(A) \bmod p_n = \mathrm{per}(A)$. This strategy works due to our Thm. 4.4 stating that a system of integer polynomial equations, which is solvable over the complex numbers, is also solvable over many finite fields $\mathbb{F}_p$.

A second ingredient of the proof is Cor. 2.25 about the efficient parallelization of straight-line programs.

Our main Thm. 4.5 is inspired by similar investigations in the realm of the BSS-model, where one also tries to identify the Boolean parts of certain algebraic complexity classes. (See the various papers authored by Blum, Cucker, Grigoriev, Koiran, Shub, and Smale [66, 33, 32, 10, 30].) There also, the "elimination of constants" is the main difficulty.

Corollary 4.6 (1) *We assume (GRH). If Valiant's hypothesis were false over a field of characteristic zero, then we had*

$$\mathrm{NC}^3/\mathrm{poly} = \mathrm{P}/\mathrm{poly} = \mathrm{NP}/\mathrm{poly} = \mathrm{PH}/\mathrm{poly}$$

and $\#\mathrm{P}/\mathrm{poly} = \mathrm{FP}/\mathrm{poly}$.

(2) *If Valiant's hypothesis were false over a finite field k of characteristic p, then we had*

$$\mathrm{NC}^2/\mathrm{poly} = \mathrm{P}/\mathrm{poly} = \mathrm{NP}/\mathrm{poly} = \mathrm{Mod}_p\mathrm{NP}/\mathrm{poly} = \mathrm{PH}/\mathrm{poly} \ .$$

In both situations, the polynomial hierarchy would collapse to the second level.

In order to conclude this corollary from Thm. 4.5 over finite fields, we will need an auxiliary result (Thm. 4.10), which relates NP to counting classes. This auxiliary result is obtained by combining a technique of Valiant and Vazirani [114] with Adleman's trick [1]. The statement about the possible collapse of the polynomial hierarchy then follows from a well-known result of Karp and Lipton [61]. We remark that in [112], Valiant had already drawn the conclusion $\mathrm{P}/\mathrm{poly} \neq \mathrm{NP}/\mathrm{poly} \Rightarrow \mathrm{VP} \neq \mathrm{VNP}$ over $\mathbb{F}_2$.

Problem 4.2 Can similar conclusions be drawn for infinite fields of positive characteristic?

4.3   Review of Discrete Complexity Classes

We briefly recall the definition of some discrete complexity classes. For more details, we refer to the survey by Johnson [55] and the textbook by Papadimitriou [87].

In the sequel (φ_n) will stand for a sequence of Boolean functions

$$\varphi_n \colon \{0,1\}^n \to \{0,1\}^{m(n)}, x \mapsto (\varphi_{n,1}(x), \ldots, \varphi_{n,m(n)}(x)) \ .$$

This defines the function $\varphi \colon \{0,1\}^* \to \{0,1\}^*$, $x \in \{0,1\}^n \mapsto \varphi_n(x)$, having the property that the length of $\varphi(x)$ depends only on the length $n = |x|$ of x. We will call such φ *string functions* and identify them with (φ_n).

In what follows, i stands for an integer and p is a prime number.

- FP denotes the class of all string functions which can be computed by a polynomial time Turing machine.

- FNC^i is the class of string functions which can be computed by logspace uniform families of Boolean circuits of polynomial size and depth $O(\log^i n)$.

- P and NC^i are the complexity classes of languages corresponding to FP and FNC^i. (That is, they are obtained by restricting attention to string functions of the form $\varphi \colon \{0,1\}^* \to \{0,1\}$.)

- #P consists of the functions $\phi \colon \{0,1\}^* \to \mathbb{N}$ of the form $\phi(x) = \#\{y \mid (x,y) \in R\}$, where $R \subseteq \{0,1\}^* \times \{0,1\}^*$ is a balanced, polynomial time decidable string relation.

- NP consists of the languages $\{x \in \{0,1\}^* \mid \phi(x) \geq 1\}$, where ϕ is a function in #P.

- $\#_p\mathrm{P}$ consists of the functions $\psi \colon \{0,1\}^* \to \mathbb{F}_p$, $x \mapsto \phi(x) \bmod p$, where ϕ is a function in #P.

- $\mathrm{Mod}_p\mathrm{NP}$ is the set of languages $\{x \in \{0,1\}^* \mid \phi(x) \equiv 1 \bmod p\}$, where ϕ is a function in #P.

- PH denotes the class of languages in the polynomial hierarchy.

- $\mathrm{FP}^{\#\mathrm{P}}$ is the class of string functions computable in polynomial time using an oracle in #P.

We remark that for $p = 2$, the classes $\#_p\mathrm{P}$ as well as $\mathrm{Mod}_p\mathrm{NP}$ coincide with the class parity polynomial time $\oplus\mathrm{P}$.

In the sequel, it will be useful to consider the elements of #P and $\#_p\mathrm{P}$ as string functions by identifying natural numbers or elements of $\mathbb{F}_p$ with their binary encoding. More specifically, a string function (φ_n) will be considered as an element of #P iff the map $\{0,1\}^* \to \mathbb{N}$ sending $x \in \{0,1\}^n$

to $\sum_{i=1}^{m(n)} \varphi_{n,i}(x)2^{i-1}$ is contained in #P. Similarly, (φ_n) is considered as an element of $\#_p\mathrm{P}$ iff $m(n) = \lfloor \log p \rfloor + 1 =: m$ for all n and the map

$$\{0,1\}^* \to \mathbb{F}_p, \; x \mapsto \sum_{i=1}^{m} \varphi_{|x|,i}(x)2^{i-1} \bmod p$$

is contained in $\#_p\mathrm{P}$. Using this identification, we have the obvious chain of inclusions

$$\mathrm{FNC}^1 \subseteq \mathrm{FNC}^2 \subseteq \ldots \subseteq \mathrm{FP} \subseteq \#\mathrm{P} \; .$$

For any complexity class $\mathcal{C}$ of string functions we may define the corresponding *nonuniform* complexity class $\mathcal{C}/\mathrm{poly}$ as follows (cf. Karp and Lipton [61]). Let $\{0,1\}^* \times \{0,1\}^* \to \{0,1\}^*, (x,y) \mapsto \langle x,y \rangle$ be the pairing function obtained by duplicating each bit of x and y and inserting 01 in between. A *polynomial advice* is a function $\alpha \colon \mathbb{N} \to \{0,1\}^*$ such that $n \mapsto |\alpha(n)|$ is p-bounded. The class $\mathcal{C}/\mathrm{poly}$ now consists of all string functions ψ of the form

$$\psi(x) = \varphi(\langle x, \alpha(|x|) \rangle) \; ,$$

where $\varphi \in \mathcal{C}$ and α is some polynomial advice function.

It is well known that $\mathrm{FP}/\mathrm{poly}$ is the class of string functions which can be computed by families of Boolean circuits of polynomial size. If we additionally require that the nth circuit has depth $O(\log^i n)$, we get the class $\mathrm{FNC}^i/\mathrm{poly}$.

A *nonuniform polynomial time reduction* from a language A to a language B is a string function $\rho \colon \{0,1\}^* \to \{0,1\}^*$ in $\mathrm{FP}/\mathrm{poly}$ such that $A = \rho^{-1}(B)$. We remark that if $\mathcal{C}$ is a class of languages closed under polynomial time reductions, then $\mathcal{C}/\mathrm{poly}$ is closed under nonuniform polynomial time reductions.

The following lemma will be useful later on.

Lemma 4.7 *Let (φ_n) be in $\mathrm{FP}/\mathrm{poly}$ and $u \colon \mathbb{N} \to \mathbb{N}$ be p-bounded such that $n < u(n)$. We set for $x \in \{0,1\}^n$*

$$\psi_n(x) := \sum_{i,e} \varphi_{u(n),i}(x_1, \ldots, x_n, e_{n+1}, \ldots, e_{u(n)})\, 2^{i-1} \; ,$$

where the sum is over all $1 \leq i \leq m(u(n))$ and $e \in \{0,1\}^{u(n)-n}$. Then (ψ_n) is in $\#\mathrm{P}/\mathrm{poly}$.

Proof. We set $\tilde{m}(n) := m(u(n))$. Consider the set R_n consisting of the tupels (x,y,z,e) in $\{0,1\}^n \times \{0,1\}^{\tilde{m}(n)} \times \{0,1\}^{\tilde{m}(n)} \times \{0,1\}^{u(n)-n}$ satisfying for some $0 \leq i < \tilde{m}(n)$:

$$
\begin{aligned}
y &= (0, \ldots, 0, 1, 0, \ldots, 0) &&\text{(the 1 at position i)} \\
z &= (z_1, \ldots, z_{i-1}, 0, \ldots, 0) &&\text{for some } z_1, \ldots, z_{i-1} \in \{0,1\} \\
\varphi_{u(n),i}(x,e) &= 1 &&.
\end{aligned}
$$

The union R of the sets R_n over all $n \in \mathbb{N}$ is obviously decidable in nonuniform polynomial time. We have

$$\#\{(y, z, e) \mid (x, y, z, e) \in R\} = \sum_i \sum_e \varphi_{u(n),i}(x, e)\, 2^{i-1} = \psi_{|x|}(x) \ .$$

Hence (ψ_n) is contained in $\#\mathrm{P}/\mathrm{poly}$. $\square$

In fact, it is not hard to see that the conclusion of the lemma is also valid if $(\varphi_n) \in \#\mathrm{P}/\mathrm{poly}$.

Finally, we give a detailed definition of the notion of Boolean parts.

Definition 4.8   (1) Let (f_n) be a p-family with $f_n \in k[X_1, \ldots, X_n]$. A string function (φ_n) is called a *Boolean part* of (f_n) if we have

- in the case char$k = 0$: $m(n) = n^{O(1)}$ and

$$\forall x \in \{0,1\}^n : f_n(x) = \sum_{i=1}^{m(n)} \varphi_{n,i}(x) 2^{i-1} \ ,$$

- in the case char$k = p > 0$: $m(n) = m := \lfloor \log p \rfloor + 1$ and

$$\forall x \in \{0,1\}^n : f_n(x) = \sum_{i=1}^{m} \varphi_{n,i}(x) 2^{i-1} \bmod p \ .$$

(2) The *Boolean part* $\mathrm{BP}(\mathrm{VP}_k)$ of VP_k is defined as the set of all Boolean parts of p-computable families over k. The Boolean part $\mathrm{BP}(\mathrm{VNP}_k)$ is defined as the set of all Boolean parts of p-definable families over k.

Remark 4.9 Let (f_n) be a p-family with $f_n \in k[X_1, \ldots, X_n]$.

(1) If char$k = p > 0$, then the family (f_n) has a Boolean part (φ_n) iff all f_n have coefficients in the prime field $\mathbb{F}_p$. In this case (φ_n) is unique. We remark that the restriction to the prime field is just for convenience and not essential in the sequel.

(2) If char$k = 0$, then the family (f_n) has a Boolean part if $f_n(x)$ are natural numbers of bitsize n^c for all $x \in \{0,1\}^n$ and some constant c. In this case, (f_n) has infinitely many Boolean parts (leading zeros).

4.4   Relating NP to Counting Classes

In order to deduce Cor. 4.6 from Thm. 4.5 over finite fields, we need the following result.

Theorem 4.10 *For a prime p we have*

$$\mathrm{NP}/\mathrm{poly} \subseteq \mathrm{Mod}_p\mathrm{NP}/\mathrm{poly} \ .$$

Proof. The problem SAT to decide for a given Boolean formula ϕ in conjunctive normal form whether it is satisfiable, is NP-complete by Cook's theorem. Let $\#\phi$ denote the number of satisfying assignments of ϕ. The problem $\mathrm{Mod}_p\mathrm{SAT}$ to find out whether $\#\phi \equiv 1 \bmod p$ for a given conjunctive normal formula (cnf, for short) is clearly in the class $\mathrm{Mod}_p\mathrm{NP}$. To prove the asserted inclusion, it is sufficient to construct a nonuniform polynomial time reduction from SAT to $\mathrm{Mod}_p\mathrm{SAT}$. (Note that $\mathrm{Mod}_p\mathrm{NP}/\mathrm{poly}$ is closed under such reductions.) Therefore, we need to find a function computable in nonuniform polynomial time, which maps a cnf ϕ to a cnf χ such that

$$(4.1) \qquad \#\phi > 0 \iff \#\chi \equiv 1 \bmod p \ .$$

The easy proof of the following lemma is left to the reader.

Lemma 4.11 *Let ϕ and ψ be cnfs in the disjoint set of variables $X_1, \ldots, X_m$ and $Y_1, \ldots, Y_n$, respectively.*

(1) $\#(\phi \wedge \psi) = \#\phi \cdot \#\psi$.

(2) *The formula $\sigma := (Z \to \phi) \wedge \bigwedge_{j=1}^n (Z \to Y_j) \wedge (\overline{Z} \to \psi) \wedge \bigwedge_{i=1}^m (\overline{Z} \to X_i)$ satisfies $\#\sigma = \#\phi + \#\psi$. It can be transformed into an equivalent cnf in polynomial time.*

(3) *From the natural number t (given in binary) we can compute in polynomial time a cnf ϕ_t satisfying $\#\phi_t = t$.*

(4) *From cnfs $\Phi_1, \ldots, \Phi_q$ and a prime p we can compute in polynomial time a cnf χ such that*

$$\#\chi = 1 + \prod_{j=1}^q (p - 1 + \#\Phi_j) \ .$$

Our proof of Thm. 4.10 heavily relies on the reduction in Valiant and Vazirani [114]. They showed that one can assign to a cnf ϕ in n variables and to a random bitstring w of length n, a cnf Φ such that

$$\#\phi = 0 \implies \#\Phi = 0 \ , \quad \#\phi > 0 \implies \mathrm{Prob}[\#\Phi \neq 1] \leq 1 - (4n)^{-1} \ .$$

Moreover, the map $(\phi, w) \mapsto \Phi$ is computable in polynomial time.

Let now a cnf ϕ with n variables and an odd natural number q be given. Choose q random bitstrings $w_1, \ldots, w_q$ of length n. Let Φ_i be the cnf assigned to ϕ and w_i by the reduction of Valiant and Vazirani. Further, let χ be the cnf corresponding to $\Phi_1, \ldots, \Phi_q$ (and the fixed prime p) according to Lemma 4.11(4). Then χ can be computed from $(\Phi, w_1, \ldots, w_q)$ in polynomial time. Moreover, we have

$$\#\phi = 0 \implies \#\chi \equiv 0 \bmod p, \quad \#\phi > 0 \implies \mathrm{Prob}[\#\chi \not\equiv 1 \bmod p] \leq (1 - (4n)^{-1})^q \ .$$

Now we proceed with Adleman's trick [1]. Let N_a be the number of cnfs of size $a \geq n$. It is easy to see that $\log N_a = a^{O(1)}$. If we choose $q = a^{O(1)}$ big enough, then

$$N_a (1 - (4n)^{-1})^{n\frac{q}{n}} \leq N_a e^{-\frac{q}{4n}} < 1 \ .$$

This implies that for each a there exist $w_1, \ldots, w_q$ such that for *all* cnfs ϕ of size a we have

$$\#\phi > 0 \Longrightarrow \#\chi \equiv 1 \bmod p \ .$$

The bitstrings $w_1, \ldots, w_q$ then serve as an advice to the cnfs of size a. This proves the claim (4.1) and thus the theorem. $\qquad\qquad\square$

We deduce now Cor. 4.6 from Thm. 4.5. Assume we had $\mathrm{VP}_k = \mathrm{VNP}_k$. For $\mathrm{char} k = 0$ we then obtain from Thm. 4.5 under (GRH) that

$$\#\mathrm{P}/\mathrm{poly} \subseteq \mathrm{BP}(\mathrm{VNP}_k) = \mathrm{BP}(\mathrm{VP}_k) \subseteq \mathrm{FNC}^3/\mathrm{poly} \subseteq \mathrm{FP}/\mathrm{poly} \subseteq \#\mathrm{P}/\mathrm{poly} \ ,$$

thus we have equality everywhere. In particular, $\mathrm{P}/\mathrm{poly} = \mathrm{NP}/\mathrm{poly}$. It is well-known that $\mathrm{P} = \mathrm{NP}$ implies $\mathrm{P} = \mathrm{PH}$ (cf. [55]). A similar argument shows that $\mathrm{P}/\mathrm{poly} = \mathrm{NP}/\mathrm{poly}$ implies $\mathrm{P}/\mathrm{poly} = \mathrm{PH}/\mathrm{poly}$. Moreover, by Karp and Lipton [61], the statement $\mathrm{P}/\mathrm{poly} = \mathrm{NP}/\mathrm{poly}$ implies the collapse of the polynomial hierarchy at the second level.

In the case of a finite field k of characteristic p we argue as follows. If $\mathrm{VP}_k = \mathrm{VNP}_k$, we get from Thm. 4.5 as above

$$\#_p\mathrm{P}/\mathrm{poly} \subseteq \mathrm{FNC}^2/\mathrm{poly} \subseteq \mathrm{FP}/\mathrm{poly} \ .$$

Switching to the corresponding classes of languages we obtain

$$\mathrm{Mod}_p\mathrm{NP}/\mathrm{poly} \subseteq \mathrm{NC}^2/\mathrm{poly} \subseteq \mathrm{P}/\mathrm{poly} \subseteq \mathrm{NP}/\mathrm{poly} \ .$$

By invoking Thm. 4.10 we see that we have equality in the above chain of inclusions. $\qquad\qquad\square$

4.5   A Bound on the Heights

Let (S) be a system of equations as in Sect. 4.2. The first ingredient of the proof of Thm. 4.4 is the following estimate of the heights of solutions of the system (S). It is based on a result in Krick and Pardo [68, 69], which itself heavily relies on the techniques in Giusti and Heintz [45].

Theorem 4.12 *The system (S) has a solution $x = (x_i)$ of the form $x_i = \theta^{-1} v_i(y)$. Here θ is a positive integer, $v_i \in \mathbb{Z}[Y]$ is an integer polynomial, and y is an algebraic number with primitive integer minimal polynomial $g \in \mathbb{Z}[Y]$ such that*

$$\max\{\deg g, \deg v_i\} = d^{O(n)}, \quad \max\{\log \theta, \log \mathrm{wt}(g), \log \mathrm{wt}(v_i)\} = d^{O(n)} \log w \ .$$

For the proof of this result we require several lemmas.

Lemma 4.13 *Let $U_1, \ldots, U_t$ be proper linear subspaces of $\mathbb{Q}^{s+1}$ and let $\Sigma_s(t) = \{a \in \mathbb{N}^{s+1} \mid \sum_{\sigma=0}^{s} a_\sigma = t\}$ be the set of lattice points of an s-dimensional simplex. Then we have*

$$\Sigma_s(t) \not\subseteq U_1 \cup U_2 \cup \ldots \cup U_t \ .$$

Proof. We proceed by induction on $s+t$. The start is clear. For the induction step we distinguish two cases. Assume first that $\mathbb{Q}^s \times \{0\}$ is not contained in any of the U_τ. Then the U_τ' defined by $U_\tau' \times \{0\} = U_\tau \cap (\mathbb{Q}^s \times \{0\})$ are proper linear subspaces of $\mathbb{Q}^s$, and we have by the induction hypothesis

$$\Sigma_{s-1}(t) \not\subseteq U_1' \cup \ldots \cup U_t' \ ,$$

hence $\Sigma_s(t) \not\subseteq U_1 \cup \ldots \cup U_t$. If $\mathbb{Q}^s \times \{0\}$ equals one of the U_τ, say U_t, then we get

$$\Sigma_s(t-1) \simeq \Sigma_s(t) \cap \{a_s > 0\} \not\subseteq U_1 \cup \ldots \cup U_{t-1}$$

by the induction hypothesis. Therefore $\Sigma_s(t) \not\subseteq U_1 \cup \ldots \cup U_t$. $\square$

We remark that the lemma is optimal in the sense that $t+1$ proper linear subspaces are sufficient to cover $\Sigma_s(t)$.

In what follows, let δ denote the codimension of the zeroset $Z(f_1, \ldots, f_s)$ in $\mathbb{C}^n$.

Lemma 4.14 *There exist $a_{i\sigma} \in \mathbb{N}$ for $1 \leq i \leq \delta$, $1 \leq \sigma \leq s$, such that the zeroset of the polynomials*

$$F_i = \sum_{\sigma=1}^{s} a_{i\sigma} f_\sigma, \ \ 1 \leq i \leq \delta \ ,$$

has codimension δ in $\mathbb{C}^n$, and such that $\sum_{\sigma=1}^{s} a_{i\sigma} \leq d^{\delta-1}$.

Proof. Assume we have already constructed $F_1, \ldots, F_r$ such that the codimension of the zeroset W of these polynomials equals $r < \delta$. Let $C_1, \ldots, C_t$ be the irreducible components of W of (maximal) dimension $n - r$. By Bézout's inequality (cf. [21, 8.28]), we know that $t \leq \sum_{\tau=1}^{t} \deg C_\tau \leq d^r \leq d^{\delta-1}$. The linear spaces

$$U_\tau := \left\{ a \in \mathbb{Q}^s \,\middle|\, \sum_{\sigma=1}^{s} a_\sigma f_\sigma \text{ vanishes on } C_\tau \right\}$$

are proper subspaces of $\mathbb{Q}^s$, since otherwise C_τ would be contained in the zeroset of $f_1, \ldots, f_s$, which implied the contradiction $r = \operatorname{codim} C_\tau \geq \delta$. By Lemma 4.13, there exists $a \in \mathbb{N}^s$ satisfying $a \notin U_1 \cup \ldots \cup U_t$ and $\sum_{\sigma=1}^{s} a_\sigma = t$. We define $F_{r+1} := \sum_{\sigma=1}^{s} a_\sigma f_\sigma$ and obtain $\operatorname{codim} Z(F_1, \ldots, F_{r+1}) = r + 1$ as desired. $\square$

We note that Lemma 4.14 implies the following fact: if $C_1, \ldots, C_t$ are the irreducible components of $Z(f_1, \ldots, f_s)$ of maximal dimension, then

$$(4.2) \qquad \sum_{\tau=1}^{t} \deg C_\tau \le d^\delta \ .$$

Lemma 4.15 *There exist affine linear polynomials $g_1, \ldots, g_{n-\delta}$ such that the zeroset $Z(f_1, \ldots, f_s, g_1, \ldots, g_{n-\delta})$ is zero-dimensional and* $\mathrm{wt}(g_\mu) \le d^{n-1} + 1$.

Proof. Assume we have already constructed $g_1, \ldots, g_m$, $m < n - \delta$, such that $W := Z(f_1, \ldots, f_s, g_1, \ldots, g_m)$ has codimension $\delta + m$. Let $C_1, \ldots, C_t$ be the irreducible components of W of maximal dimension. By observation (4.2) we know that $\sum_{\tau=1}^{t} \deg C_\tau \le d^{\delta+m}$.

We need to prove the existence of a "small" nonzero $a \in \mathbb{N}^{n+1}$ such that the affine hyperplane $H := Z(a_0 + \sum_{i=1}^{n} a_i X_i)$ has a nonempty intersection with W, but does not contain any of the C_τ. For this, it is convenient to take a projective point of view. Think of $\mathbb{C}^n$ as being embedded in $\mathbb{P}^n$ and let $\mathcal{H}$ be the Grassmannian consisting of the hyperplanes in $\mathbb{P}^n$. Note that $\mathcal{H}$ can be naturally identified with $\mathbb{P}^n$ itself.

There is a a curve K contained in C_1 such that $\deg K \le \deg C_1$. (Just intersect C_1 with an affine subspace of complementary dimension which is in general position, and apply Bézout's inequality.) Let $\overline{K}$ be the projective closure of K. Then we have $\overline{K} = K \cup K_\infty$, where K_∞ is the finite set of points of $\overline{K}$ at infinity. By Bézout's inequality, we have $|K_\infty| \le \deg K$. For $p \in \mathbb{P}^n$ consider the hyperplanes

$$\mathcal{V}_p := \{H \in \mathcal{H} \mid p \in H\}$$

in $\mathcal{H}$. If some $H \in \mathcal{H}$ does not intersect W, then it does not intersect the affine part K of $\overline{K}$, hence it must intersect $\overline{K}$ in one of its points at infinity. To put it differently: any hyperplane H lying in none of the $\mathcal{V}_p$, $p \in K_\infty$, must necessarily intersect W.

Consider now the following proper linear subspaces of $\mathcal{H}$

$$\mathcal{U}_\tau := \{H \in \mathcal{H} \mid C_\tau \subseteq H\} \ .$$

Lemma 4.13 shows the existence of some $a \in \mathbb{N}^{n+1}$ such that the hyperplane $H = Z(\sum_{i=0}^{n} a_i X_i)$ satisfies

$$H \notin \bigcup_{\tau=1}^{t} \mathcal{U}_\tau \cup \bigcup_{p \in K_\infty} \mathcal{V}_p \ ,$$

and $\sum_{i=0}^{n} a_i = t + |K_\infty|$. (To obtain this, interpret the $\mathcal{U}_\tau$ and $\mathcal{V}_p$ as linear subspaces of $\mathbb{C}^{n+1}$.) Since

$$\begin{aligned} t + |K_\infty| &\le t + \deg C_1 \le \deg C_1 + \ldots \deg C_t + 1 \\ &\le d^{\delta+m} + 1 \le d^{n-1} + 1 \ , \end{aligned}$$

this proves the assertion. $\qquad\qquad\square$

Recall that the weight $\mathrm{wt}(f)$ of a polynomial $f \in \mathbb{Z}[X_1, \ldots, X_n]$ is the sum of the absolute values of its coefficients. It is easy to check that the weight is subadditive and submultiplicative:

$$\mathrm{wt}(f + g) \leq \mathrm{wt}(f) + \mathrm{wt}(g), \quad \mathrm{wt}(f \cdot g) \leq \mathrm{wt}(f) \cdot \mathrm{wt}(g) \ .$$

Let Γ be a straight-line program using the operations $+, -, *$. Each node ρ in the acyclic multigraph associated with Γ has a *multiplicative depth* $d_*(\rho)$, which is defined similarly as the usual depth, but only multiplication nodes are counted. Analogously, we define the *additive depth* $d_+(\rho)$ (counting only additions and subtractions). The multiplicative and additive depth of Γ are defined as the maximum of the numbers $d_*(\rho)$ and $d_+(\rho)$ taken over all ρ, respectively.

Lemma 4.16 *Let $f \in \mathbb{Z}[X_1, \ldots, X_n]$ be computed by a straight-line program Γ from the variables X_i and integers of absolute value at most $b \geq 2$. Then we have*

$$\deg f \leq 2^{d_*}, \quad \log \mathrm{wt}(f) \leq (d_+ + 1)\, 2^{d_*} \log b \ ,$$

where d_* and d_+ denote the multiplicative and additive depth of Γ.

Proof. Let (g_ρ) be the result sequence of Γ. We prove by induction on ρ that

$$\log \mathrm{wt}(g_\rho) \leq (d_+(\rho) + 1)\, 2^{d_*(\rho)} \log b \ .$$

If $g_\rho = g_i \cdot g_j$ with $i, j < \rho$, then $d_*(\rho) = 1 + \max\{d_*(i), d_*(j)\}$ and $d_+(\rho) = \max\{d_+(i), d_+(j)\}$. We thus have

$$
\begin{aligned}
\log \mathrm{wt}(g_\rho) \ &\leq \ \log \mathrm{wt}(g_i) + \log \mathrm{wt}(g_j) \\
&\leq \ 2\,(d_+(\rho) + 1) \max\{2^{d_*(i)}, 2^{d_*(j)}\} \log b \\
&= \ (d_+(\rho) + 1)\, 2^{d_*(\rho)} \log b \ .
\end{aligned}
$$

If $g_\rho = g_i + g_j$, $i, j < \rho$, then $d_*(\rho) = \max\{d_*(i), d_*(j)\}$ and $d_+(\rho) = 1 + \max\{d_+(i), d_+(j)\}$. Hence

$$
\begin{aligned}
\log \mathrm{wt}(g_\rho) \ &\leq \ 1 + \max\{\log \mathrm{wt}(g_i), \log \mathrm{wt}(g_j)\} \\
&\leq \ 1 + d_+(\rho)\, 2^{d_*(\rho)} \log b \leq (d_+(\rho) + 1)\, 2^{d_*(\rho)} \log b \ .
\end{aligned}
$$

The induction start is clear, and the proof of the degree estimate is left to the reader. $\qquad\square$

We are now ready for the proof of Thm. 4.12.

Proof. (of Thm. 4.12) By Lemma 4.15, we can add to the system (S) suitable linear equations of weight at most d^n, such that the resulting system becomes zero-dimensional. By Lemma 4.14, we obtain n integer polynomials $F_1, \ldots, F_n$

satisfying $\deg F_i \leq d$, $\mathrm{wt}(F_i) \leq w d^{2n}$, and such that $V := Z(F_1, \ldots, F_n)$ is zero-dimensional and contains a solution of the original system (S).

We can now apply Prop. 27 of [69], which claims the existence of an integer linear form $\gamma = \gamma_1 X_1 + \ldots + \gamma_n X_n$, of a positive integer θ, and of univariate integer polynomials $v_i \in \mathbb{Z}[Y]$, $1 \leq i \leq n$, such that the following holds:

(1) The linear form γ separates the points of V.

(2) We have $\theta x_i = v_i(\gamma(x))$ for all $x = (x_1, \ldots, x_n) \in V$.

(3) $\mathrm{wt}(\gamma) = d^{O(n)}$.

(4) Both θ and the coefficients of the v_i are integer polynomials in the coefficients of $F_1, \ldots, F_n$, and they can be evaluated from these coefficients and the constant 1 by a division-free straight-line program of size $d^{O(n)}$ and multiplicative depth $O(n \log d)$.

(The original statement in [69, Prop. 27] is about nonscalar straight-line programs, but it is easily seen to be equivalent to our formulation.)

Taking into account that the coefficients of F_i are bounded by $w d^{2n}$, we conclude with Lemma 4.16 that

$$\deg v_i = d^{O(n)}, \quad \max\{\log \theta, \log \mathrm{wt}(v_i)\} = d^{O(n)} \log w \ .$$

Choose $x \in V$ and set $y := \gamma(x)$. Let g be a minimal polynomial of y, which we moreover assume to be a primitive integer polynomial. We need to show the desired estimates of the degree and weight of g. Consider the univariate integer polynomials

$$G_i(Y) := \theta^d F_i(\theta^{-1} v_1(Y), \ldots, \theta^{-1} v_n(Y)) \ .$$

We may assume that at least one of the G_i is nonzero, since otherwise all v_i are constant, and we are done already. Note that $G_i(y) = 0$ for all i. Hence g is a divisor of G_i in $\mathbb{Z}[Y]$. Now we are going to bound the degree and weight of G_i. Write $F_i = \sum_\mu c_\mu X_1^{\mu_1} \cdots X_n^{\mu_n}$. Then we have

$$G_i = \sum_\mu c_\mu \theta^{d - |\mu|} v_1^{\mu_1} \cdots v_n^{\mu_n} \ ,$$

hence $\deg G_i \leq \deg F_i \cdot \max_i \deg v_i = d^{O(n)}$. Using the subadditivity and submultiplicativity of the weight we infer that $\mathrm{wt}(G_i) \leq \theta^d \mathrm{wt}(F_i)(\max_i \mathrm{wt}(v_i))^d$, which implies $\log \mathrm{wt}(G_i) = d^{O(n)} \log w$.

The bound in Mignotte [85] on the coefficients of divisors of polynomials

$$\mathrm{wt}(g) \leq \frac{\mathrm{lc}(g)}{\mathrm{lc}(G_i)} 2^{\deg g} \|G_i\|$$

implies the desired estimate of $\mathrm{wt}(g)$. (Here, $\mathrm{lc}(g)$ denotes the leading coefficient of g and $\|G_i\|$ is the L_2-norm of the coefficient vector of G_i.) $\qquad \square$

Remark 4.17 We adopt the notation of Thm. 4.12. Let $\overline{y}$ be a root of g in $\mathbb{F}_p$ and assume that p is not a factor of θ. Consider the integer polynomials

$$h_i(Y) := \theta^d f_i(\theta^{-1} v_1(Y), \ldots, \theta^{-1} v_n(Y)) \ .$$

As $(\theta^{-1} v_i(y))_i$ is a solution of (S), we have $h_i(y) = 0$. Therefore, the minimal polynomial g is a factor of h_i in $\mathbb{Q}[Y]$. Since we assume g to be primitive, it is even a factor of h_i in $\mathbb{Z}[Y]$. By our assumption, we have $g(\overline{y}) = 0$ in $\mathbb{F}_p$. Hence $h_i(\overline{y}) = 0$ and we conclude that $(\theta^{-1} v_i(\overline{y})_i$ is a solution of (S) over $\mathbb{F}_p$.

So it remains to investigate the distribution of the roots mod p of univariate polynomials. This will be done in the next section.

4.6 Roots of Univariate Polynomials Modulo a Prime

Let g be an irreducible univariate integer polynomial of degree d. It is well known that there is a bijection between the irreducible factors of g modulo p and the primes of the number field $K = \mathbb{Q}[Y]/(g)$ lying over p, provided p is not a divisor of the discriminant Δ of g (cf. [79]). Under this bijection, the roots of g modulo p correspond to the primes of degree one lying over p. Let $\pi_K(x)$ denote the number of primes of K having norm at most x, and $\mathrm{li}(x) = \int_2^x du/\ln u \sim x/\ln x$ be the logarithmic integral. It is known that under the generalized Riemann hypothesis (GRH) the following effective version of the "prime number theorem" for number fields is true (Weinberger [116], see also Lagarias and Odlyzko [71]):

$$(4.3) \qquad |\pi_K(x) - \mathrm{li}(x)| = O(x^{1/2} \log(|\Delta| x^d)) \ .$$

(The constant implicit in the O-term does not depend on g.) Of course, this contains an effective version of the usual prime number theorem ($K = \mathbb{Q}$) as a special case. We recall that the *generalized Riemann hypothesis* (GRH) claims that all complex roots $s + it$ of the zeta function ζ_K of K in the critical strip $0 \le s \le 1$ satisfy $s = 1/2$.

A prime ideal of K having norm $\le x$ and degree > 1 lies over a rational prime $p \le x^{1/2}$. Hence there are at most $d x^{1/2}$ such primes of K. Moreover, there are at most $\log|\Delta|$ prime factors of Δ. By taking into account these considerations, and using also the effective prime number theorem for $\mathbb{Q}$, one can easily deduce from (4.3) the following result (cf. Weinberger [117]).

Theorem 4.18 *Let $g \in \mathbb{Z}[Y]$ be irreducible with degree d and discriminant Δ. Then the total number $N_g(x)$ of roots of g modulo all the primes up to x satisfies*

$$|N_g(x) - \pi(x)| = O(x^{1/2} \log(|\Delta| x^d) + d \log |\Delta|) \ ,$$

provided (GRH) is true.

Let $\pi_g(x)$ stand for the number of primes $p \leq x$ such that g has a root modulo p. It is obvious that $N_g(x) \leq d\,\pi_g(x)$. Moreover, if w is an upper bound on the weight of g, then we have the estimate $\log|\Delta| = O(d\log(dw))$ for the discriminant Δ. Taking these observations into account, we obtain the following corollary.

Corollary 4.19 *For all irreducible univariate polynomials $g \in \mathbb{Z}[Y]$ of degree d and weight w we have*

$$\pi_g(x) \geq \frac{\pi(x)}{d} - O(x^{1/2}\log(dwx) + d\log(dw)) \ ,$$

provided (GRH) is true.

The proof of Thm. 4.4 follows now easily by combining Thm. 4.12, Remark 4.17, and Cor. 4.19.

4.7   Proof of Thm. 4.5

(A) We first discuss the more complicated case where $\mathrm{char}\,k = 0$.

(A1) $\mathrm{FNC}^1/\mathrm{poly} \subseteq \mathrm{BP}(\mathrm{VP}_k)$:

Let (C_n) be a sequence of Boolean circuits of size $s_n = n^{O(1)}$ and depth $d_n = O(\log n)$ and φ_n be the function computed by C_n. Using the relations

$$\forall x, y \in \{0,1\} : x \wedge y = x \cdot y, \ \neg x = 1 - x, \ x \vee y = x + y - x \cdot y$$

we can simulate C_n by a straight-line program Γ_n of size at most $3s_n$ and depth bounded by at most $2d_n$. If $g_{n,i}$ are the polynomials computed by Γ_n, we have $g_{n,i}(x) = \varphi_{n,i}(x)$ for all x in $\{0,1\}^n$. Thus (φ_n) is a Boolean part of (f_n) defined by $f_n := \sum_{i=1}^{m(n)} g_{n,i}2^{i-1}$. The family (f_n) is p-computable, since its degree is growing at most polynomially due to

$$\deg g_n \leq 2^{\mathrm{depth}\,\Gamma_n} \leq n^{O(1)} \ .$$

(A2) $\#\mathrm{P}/\mathrm{poly} \subseteq \mathrm{BP}(\mathrm{VNP}_k)$:

Let (φ_n) be a string function in $\#\mathrm{P}$ and set $\phi(x) := \sum_i \varphi_{n,i}(x)2^{i-1}$ for x in $\{0,1\}^n$. By our definition $\phi \colon \{0,1\}^* \to \mathbb{N}$ is in $\#\mathrm{P}$. As in the proof of Valiant's criterion (Prop. 2.20), we see that there is a p-computable family (p_n), such that the p-definable family (g_n) defined by

$$g_n(X) := \sum_{y \in \{0,1\}^{m(n)}} p_n(X, y) \ ,$$

satisfies $g_n(x) = \phi(x)$ for all x in $\{0,1\}^n$. Therefore, (φ_n) is a Boolean part of (g_n), which shows that $\#\mathrm{P} \subseteq \mathrm{BP}(\mathrm{VNP}_k)$.

Now let (ψ_n) be in #P/poly. By definition, there is some (φ_n) in #P and some polynomial advice function α such that for all $x \in \{0,1\}^n$

$$\psi_n(x) = \varphi_{t(n)}(\langle x, \alpha(n)\rangle) \ ,$$

where $t(n) = 2n + 2|\alpha(n)| + 2$ is the length of $\langle x, \alpha(n)\rangle$. By the reasoning before, there exists $(g_n) \in \mathrm{VNP}_k$ such that $g_n(x) = \sum_i \varphi_{n,i}(x) 2^{i-1}$ for $x \in \{0,1\}^n$. Now define the polynomial $h_n(X) := g_{t(n)}(\langle X, \alpha(n)\rangle)$ (the pairing has the obvious meaning). (h_n) is p-definable as it is a p-projection of (g_n). On the other hand, we have for all $x \in \{0,1\}^n$

$$h_n(x) = g_{t(n)}(\langle x, \alpha(n)\rangle) = \sum_i \varphi_{t(n),i}(\langle x, \alpha(n)\rangle)\, 2^{i-1} = \sum_i \psi_{n,i}(x) 2^{i-1} \ .$$

Therefore, (ψ_n) is a Boolean part of (h_n) and hence $(\psi_n) \in \mathrm{BP}(\mathrm{VNP}_k)$.

(A3) $\mathrm{BP}(\mathrm{VP}_k) \subseteq \mathrm{FNC}^3/\mathrm{poly}$:

Assume $(f_n) \in \mathrm{VP}_k$ has a Boolean part. Hence there is some p-bounded function $t \colon \mathbb{N} \to \mathbb{N}$ such that $f_n(x) < 2^{t(n)}$ for $x \in \{0,1\}^n$. By Cor. 2.25 there is for each n some straight-line program Γ_n of size $n^{O(1)}$ and depth $O(\log^2 n)$ which computes f_n from $X_1, \ldots, X_n$ and *constants* $y_1, \ldots, y_{m(n)}$ in k. (In particular, $m(n) = n^{O(1)}$.) Replace the y_j by indeterminates Y_j and let $F_n(X, Y)$ denote the integer polynomial computed by Γ_n from the X_i and Y_j. We conclude from Lemma 4.16 that

$$\deg F_n \le 2^{O(\log^2 n)}, \ \log \mathrm{wt}(F_n) \le 2^{O(\log^2 n)} \ .$$

Obviously, $F_n(X, y) = f_n$, hence the system of equations

$$(S_n) \qquad\qquad F_n(x, Y) - f_n(x) = 0 \quad \text{for all } x \in \{0,1\}^n$$

has a solution $y \in k^{m(n)}$. By the Nullstellensatz, the system (S_n) is also solvable over the algebraic closure of $\mathbb{Q}$ and thus over $\mathbb{C}$.

Note that $\mathrm{wt}(F_n(x, Y)) \le \mathrm{wt}(F_n)$ for $x \in \{0,1\}^n$. Hence the degree as well as the logarithm of the weight of the polynomials in (S_n) are bounded by $2^{O(\log^2 n)}$. Thm. 4.4 implies that for a suitable constant $c > 0$

$$\pi_{S_n}(2^{n^c}) > 2^{t(n)}$$

for sufficiently large n. Therefore, there is some prime number p_n satisfying $t(n) < \log p_n \le n^c$ such that (S_n) is solvable over $\mathbb{F}_{p_n}$. Let $\overline{y}_n \in \mathbb{F}_{p_n}^{m(n)}$ denote such a solution.

We remark that the addition and multiplication in $\mathbb{F}_p$ can be performed by (uniform) Boolean circuits of size $(\log p)^{O(1)}$ and depth $O(\log \log p)$ (cf. Karp and Ramachandran [62, §4.2.2]). Using this, we see that we can simulate the straight-line program Γ_n in $\mathbb{F}_{p_n}$ with constants $\overline{y}_n$ by a Boolean circuit C_n satisfying

$$
\begin{aligned}
\text{size}(C_n) &= (\log p_n)^{O(1)} \cdot \text{size}(\Gamma_n) &= n^{O(1)} \\
\text{depth}(C_n) &= O(\log \log p_n \cdot \text{depth}(\Gamma_n)) &= O(\log^3 n) \ .
\end{aligned}
$$

The circuit C_n computes a bit representation of $F_n(x, \overline{y}_n) \in \mathbb{F}_{p_n}$ on input $x \in \{0, 1\}^n$. (We can think of p_n and $\overline{y}_n$ as being "hard-wired" in C_n.) Since

$$
f_n(x) \bmod p_n = F_n(x, \overline{y}_n)
$$

and $f_n(x) < 2^{t(n)} \le p_n$, C_n actually computes a bit representation of $f_n(x)$.

(A4) $\text{BP}(\text{VNP}_k) \subseteq \text{FP}^{\#\text{P}}/\text{poly}$:

Assume $(f_n) \in \text{VNP}_k$ has a Boolean part. By definition, there is a p-computable family (g_n) and a p-bounded function $u: \mathbb{N} \to \mathbb{N}$ such that

$$
(4.4) \qquad f_n = \sum_{e \in \{0,1\}^{u(n)-n}} g_{u(n)}(X_1, \ldots, X_n, e_{n+1}, \ldots, e_{u(n)}) \ .
$$

Under the additional assumption that also (g_n) has a Boolean part (φ_n), we could easily finish the argumentation. By the inclusion (A3) just proved before, we had $(\varphi_n) \in \text{FNC}^3/\text{poly} \subseteq \text{FP}/\text{poly}$. Therefore, we would have for all $x \in \{0, 1\}^n$

$$
f_n(x) = \sum_e g_{u(n)}(x, e) = \sum_{e,i} \varphi_{u(n),i}(x, e) 2^{i-1}
$$

and Lemma 4.7 would imply that the map $\{0, 1\}^* \to \mathbb{N}, x \mapsto f_{|x|}(x)$ is in $\#\text{P}/\text{poly}$.

In the general situation, we may argue similarly as in the proof of (A3). Let $t: \mathbb{N} \to \mathbb{N}$ be p-bounded such that $f_n(x) < 2^{t(n)}$ for $x \in \{0, 1\}^n$. For each n there is some straight-line program Γ_n of size $n^{O(1)}$ and depth $O(\log^2 n)$ which computes g_n from indeterminates $X_1, \ldots, X_n, E_{n+1}, \ldots, E_{u(n)}$ and constants $y_1, \ldots, y_{m(n)}$ in k. Let $G_n(X, E, Y)$ be the integer polynomial computed by Γ_n if we replace the y_j by indeterminates Y_j. It is clear that $g_n = G_n(X, E, y)$. We will apply Thm. 4.4 to the system

$$
(S_n') \qquad \sum_{e \in \{0,1\}^{u(n)-n}} G_{u(n)}(x, e, Y) \ - \ f_n(x) \ = \ 0 \quad \text{for all } x \in \{0, 1\}^n \ ,
$$

which is solvable over $\mathbb{C}$. Note that

$$
\text{wt}\Big(\sum_e G_{u(n)}(x, e, Y)\Big) \le \sum_e \text{wt}(G_{u(n)}) \le 2^{2^{O(\log^2 n)}} \ .
$$

Hence the degree as well as the logarithm of the weight of the polynomials in (S_n') are bounded by $2^{O(\log^2 n)}$. By Thm. 4.4 there is some prime number $p_n \ge 2^{t(n)}$ of bitsize $n^{O(1)}$ such that (S_n') has a solution $\overline{y}_n \in \mathbb{F}_{p_n}^{m(n)}$.

As in (A3), we can construct for each n a Boolean circuit C_n of size $n^{O(1)}$ and depth $O(\log^3 n)$ which computes from $(x, e) \in \{0,1\}^{u(n)}$ the bit representation of a natural number $\varphi_{u(n)}(x, e)$ such that

$$\psi_n(x) := \sum_{e \in \{0,1\}^{u(n)-n}} \varphi_{u(n)}(x, e)$$

satisfies $\psi_n(x) \bmod p_n = f_n(x)$. The map $\psi \colon \{0,1\}^* \to \mathbb{N}, x \mapsto \psi_{|x|}(x)$ is in #P/poly by Lemma 4.7. Hence $\{0,1\}^* \to \mathbb{N}, x \mapsto f_{|x|}(x)$ is in $\mathrm{FP}^{\#\mathrm{P}}$/poly.

(B) Let now $k = \mathbb{F}_{p^e}$ be a finite field with p^e elements. We can represent the elements of k by bit vectors using the isomorphism $k \simeq \mathbb{F}_p[T]/(h)$ for some irreducible polynomial h over $\mathbb{F}_p$ of degree e. The arithmetic in k can be very efficiently simulated by Boolean circuits; however, for our purposes, any simulation will do.

Our proofs for (A1) and (A2) can be immediately translated to show the inclusions FNC^1/poly $\subseteq$ BP(VP$_k$) and #$_p$P/poly $\subseteq$ BP(VNP$_k$).

To prove BP(VP$_k$) $\subseteq$ FNC^2/poly we start as in (A3) with straight-line programs Γ_n of size $n^{O(1)}$ and depth $O(\log^2 n)$ using constants in k. As k is finite, Γ_n can be directly simulated by Boolean circuits of size $n^{O(1)}$ and depth $O(\log^2 n)$.

In order to show BP(VNP$_k$) $\subseteq$ #$_p$P we assume that $f_n \in \mathbb{F}_p[X_1, \ldots, X_n]$ has a representation as in (4.4) with some $(g_n) \in$ VP$_k$. Let $b_1 = 1, b_2, \ldots, b_e$ be a basis of k over $\mathbb{F}_p$ and write $g_n(x, e) = \sum_i g_{n,i}(x, e) b_i$ with $g_{n,i}(x, e) \in \mathbb{F}_p$. It is easy to see that $f_n(x) = \sum_e g_{n,1}(x, e)$. A Boolean circuit of size $n^{O(1)}$ can be constructed which computes a bit representation of $g_{n,1}(x, e)$ from x, e. Lemma 4.7 shows now that $x \mapsto f_{|x|}(x)$ is in #P/poly. $\qquad\square$

5

The Structure
of Valiant's Complexity Classes

We further develop our theory in the spirit of structural complexity and obtain analogues of well-known results by Baker, Gill, and Solovay, Ladner, and Schöning.

We show that if Valiant's hypothesis is true, then there is a p-definable family, which is neither p-computable nor VNP-complete. More generally, we define the posets of p-degrees and c-degrees of p-definable families and prove that any countable poset can be embedded in either of them, provided Valiant's hypothesis is true. Moreover, we establish the existence of minimal pairs for VP in VNP.

Over finite fields, we give a *specific* example of a family of polynomials which is neither VNP-complete nor p-computable, provided the polynomial hierarchy does not collapse.

We define relativized complexity classes VP^h and VNP^h and construct complete families in these classes. Moreover, we prove that there is a p-family h satisfying $\mathrm{VP}^h = \mathrm{VNP}^h$.

The results of this chapter are taken from Bürgisser [20].

5.1 Outline and Comparison with Previous Work

Our goal is to further develop Valiant's approach along the lines of discrete structural complexity theory.

We show that if Valiant's hypothesis is true, then, over any field, there is a p-definable family which is neither p-computable nor VNP-complete. A similar result due to Ladner [70] in the classical P-NP-setting is well-known. Ladner's proof is a diagonalization argument based on an effective enumeration of all polynomial time Turing machines. However, over uncountable structures, this approach causes problems. Malajovich and Meer [78] carried over Ladner's theorem to the setting of the BSS-model over the complex numbers by employing a transfer principle due to Blum et al. [10], which allows a reduction to the countable field of algebraic numbers. The corresponding question over the reals is still open, but it is known to be true under a nonuniformity assumption, cf. Ben-David et al. [7]. One of the reasons our proof works over any field is the nonuniformity of Valiant's model. (For a detailed treatment of these questions in a general model-theoretic context see Chapuis and Koiran [23].)

In [96] Schöning found a powerful and uniform technique for proving the existence of certain "diagonal" recursive sets. We develop a similar technique adapted to Valiant's setting. In this framework, the essence of enumeration and diagonalization arguments can be neatly captured by our notion of a σ-*limit set*, which serves as a substitute for the recursively presentable classes in Schöning's approach.

In Sect. 5.2 we formalize this in a general abstract setting by studying certain compatible quasi-orders on the set $\Omega^{\mathbb{N}}$ of families in a quasi-ordered set $(\Omega, \leq)$, and by proving an abstract diagonalization theorem. Based on this theorem, we proceed in Sect. 5.3 by providing an elegant proof that any countable poset can be embedded in the poset of degrees corresponding to a compatible quasi-order. This is applied in Sect. 5.4 in Valiant's setting to an analogue of the polynomial Turing reduction (c-reduction), as well as to the p-projection. A similar result in the classical P-NP-setting for polynomial Turing or polynomial many-one degrees was stated by Ladner [70]; however, he presented a proof in a special case only. We further remark that the existence of minimal pairs for VP in VNP can be easily guaranteed by our approach. (See Landweber et al. [72] and Schöning [97] for corresponding results in the classical P-NP setting.)

A striking discovery is that we can describe *specific* families of polynomials which are neither VNP-complete nor p-computable. In fact, the family of cut enumerators over a finite field of characteristic p has this property, provided $\mathrm{Mod}_p\mathrm{NP}$ is not contained in P/poly. (The latter condition is satisfied if the polynomial hierarchy does not collapse at the second level.) In the classical, as well as in the BSS-setting, only artificial problems are known to have such properties. This is discussed in Section 5.5.

Finally, in Sect. 5.6, we define relativized versions VP^h and VNP^h of Valiant's complexity classes with respect to a p-family h. For these, we have obtained some results in the spirit of Baker et al. [3]. (We remark that Emerson [34] has transfered such results to the BSS-model.) Over infinite fields, we can construct VP^h-complete and VNP^h-complete families with respect to p-projection. In particular, this gives a proof for the existence of VNP-complete families, which is independent of Valiant's intricate completeness proof for the permanent. Moreover, we can construct a p-family h satisfying $\mathrm{VP}^h = \mathrm{VNP}^h$. We do not know whether there exists a p-family h such that $\mathrm{VP}^h \neq \mathrm{VNP}^h$.

5.2   An Abstract Diagonalization Theorem

Let a quasi-ordered set $(\Omega, \leq)$ be fixed. Elements of the set $\Omega^{\mathbb{N}}$ of sequences in Ω will be called *(abstract) families* in the sequel. We may formally define a quasi-order $\leq_p$ on the set $\Omega^{\mathbb{N}}$ of families as in Def. 2.8. Recall that a function

$t\colon \mathbb{N} \to \mathbb{N}$ is called p-bounded from above and below iff there exists some $c > 0$ such that $n^{1/c} - c \le t(n) \le n^c + c$ for all n.

Definition 5.1   (1) We call a family $f = (f_n)$ an *(abstract) p-projection* of a family $g = (g_m)$, in symbols $f \le_p g$, iff there exists a function $t\colon \mathbb{N} \to \mathbb{N}$ which is p-bounded from above and below such that

$$\exists n_0 \ \forall n \ge n_0 : f_n \le g_{t(n)} \ .$$

(2) Two families f and g are said to be *p-equivalent* iff $f \le_p g$ and $g \le_p f$. We call the equivalence classes *(abstract) p-degrees* and denote by $\mathcal{D}_p$ the poset of all p-degrees with the partial order induced by $\le_p$.

The reader should check that $\le_p$ is indeed transitive. We will write $f <_p g$ in order to express that $f \le_p g$ but not $g \le_p f$.

Definition 5.2 The *join* $f \cup g$ of two families $f, g \in \Omega^{\mathbb{N}}$ is defined as

$$f \cup g := (f_0, g_0, f_1, g_1, f_2, g_2, \ldots) \ .$$

It is easy to see that the join of two p-degrees is well-defined and that it is the smallest upper bound of these p-degrees in $\mathcal{D}_p$. The poset $\mathcal{D}_p$ of p-degrees is thus a join-semilattice.

The following crucial notion of a *σ-limit set* captures exactly what is needed for diagonalization arguments in our framework.

Definition 5.3 By a *cylinder* in $\Omega^{\mathbb{N}}$ we shall understand a set of families of the form $F \times \Omega^{\mathbb{N}}$, where $F \subseteq \Omega^n$ for some $n \in \mathbb{N}$. A *limit of cylinders* is defined as a countable intersection of cylinders. By *σ-limit set* in $\Omega^{\mathbb{N}}$ we shall understand a countable union of limits of cylinders.

We remark that countable unions and finite intersections of σ-limit sets are again σ-limit sets. In Example 5.7 at the end of this section, we will see that the σ-limit sets are not closed under the formation of complements and countable intersections. Note that if $F_\nu \subseteq \Omega^{n_\nu}$ for $n_\nu \in \mathbb{N}$, then the cartesian product $\prod_\nu F_\nu$ is a limit of cylinders.

Lemma 5.4 $\{h \mid h \le_p g\}$ *and* $\{h \mid f \le_p h\}$ *are σ-limit sets for all $f, g \in \Omega^{\mathbb{N}}$.*

Proof.   It is convenient to use the abbreviation

$$I(c, n) := \{m \mid n^{1/c} - c \le m \le n^c + c\} \ .$$

Note that $h \le_p g$ can be expressed by the following predicate

$$\exists c > 0 \ \exists n_0 \ \forall n \ge n_0 \ \exists m : m \in I(c, n) \ \wedge \ h_n \le g_m \ ,$$

where the quantification is over natural numbers. Thus if we set

$$U(c, n) := \{u \in \Omega \mid \exists m : m \in I(c, n) \ \wedge \ u \le g_m\} \ ,$$

then we can write

$$\{h \mid h \le_p g\} = \bigcup_{c, n_0} \left(\Omega^{n_0} \times \prod_{n \ge n_0} U(c, n) \right) \ ,$$

hence $\{h \mid h \le_p g\}$ is a σ-limit set.

On the other hand, we may write $\{h \mid f \le_p h\}$ as the countable union over all c, n_0 of the following limits of cylinders

$$\bigcap_{n \ge n_0} \ \bigcup_{m \in I(n, c)} \left(\Omega^m \times \{v \in \Omega \mid f_n \le v\} \times \Omega^{\mathbb{N}} \right) \ .$$

Therefore, $\{h \mid f \le_p h\}$ is a σ-limit set. $\hfill\square$

We remark that the order $\le_p$ is compatible with the join in the following sense: for all σ-limit sets $\mathcal{F} \subseteq \Omega^{\mathbb{N}}$ the set $\{(f, g) \mid f \cup g \in \mathcal{F}\}$ is a σ-limit subset of $(\Omega \times \Omega)^{\mathbb{N}}$ via the identification $\Omega^{\mathbb{N}} \times \Omega^{\mathbb{N}} \to (\Omega \times \Omega)^{\mathbb{N}}$ sending $((f_n), (g_n))$ to $((f_n, g_n))$.

We call a family (f_n) a *finite variation* of a family (g_n) iff $f_n = g_n$ for all but finitely many n. Note that if f is a finite variation of g, then f and g are in the same p-degree. Subsets of $\Omega^{\mathbb{N}}$ which are closed under finite variation capture asymptotic properties of families (f_n) for $n \to \infty$. (In probability theory they are called tail events.)

The following abstract diagonalization theorem is inspired by Schöning's uniform diagonalization theorem [96] (see also Balcázar et al. [4]). We note that the σ-limit sets serve as a substitute for the recursively presentable classes appearing there.

Theorem 5.5 *Let $\mathcal{F}, \mathcal{G}$ be σ-limit sets of $\Omega^{\mathbb{N}}$ which are closed under finite variation. Moreover, let $f, g \in \Omega^{\mathbb{N}}$ such that $f \notin \mathcal{F}$ and $g \notin \mathcal{G}$. Then there exists $h \in \Omega^{\mathbb{N}}$ satisfying $h \le_p f \cup g$ and $h \notin \mathcal{F} \cup \mathcal{G}$.*

Proof. As $\mathcal{F}$ and $\mathcal{G}$ are σ-limit sets, we have representations $\mathcal{F} = \bigcup_i \bigcap_j \mathcal{F}_{ij}$ and $\mathcal{G} = \bigcup_i \bigcap_j \mathcal{G}_{ij}$ where $\mathcal{F}_{ij}$ and $\mathcal{G}_{ij}$ are cylinders in $\Omega^{\mathbb{N}}$. Clearly, we may assume that $\mathcal{F}_{i0} \supseteq \mathcal{F}_{i1} \supseteq \ldots$ and $\mathcal{G}_{i0} \supseteq \mathcal{G}_{i1} \supseteq \ldots$ We denote by π_n the projection $\Omega^{\mathbb{N}} \to \Omega^n, (f_\nu) \mapsto (f_0, \ldots, f_{n-1})$.

By induction, we are going to construct an infinite sequence $b_0 := 0 < a_1 < b_1 < a_2 < b_2 < \ldots$ of natural numbers such that the corresponding "mixture" h of the families f and g defined by

$$h_\nu := \begin{cases} f_\nu & \text{if } b_{s-1} \le \nu < a_s \text{ for some } s \\ g_\nu & \text{if } a_s \le \nu < b_s \text{ for some } s \end{cases}$$

satisfies for all i

$$(5.1) \qquad \pi_{a_i}(h) \notin \pi_{a_i}(\mathcal{F}_{i a_i}), \quad \pi_{b_i}(h) \notin \pi_{b_i}(\mathcal{G}_{i b_i}) \ .$$

Let us first show that the resulting h fulfills the requirements of the theorem. It is clear that $h \leq_p f \cup g$. Assume by contradiction that $h \in \mathcal{F}$. Then there exists some i such that $h \in \mathcal{F}_{ij}$ for all j. Choosing $j = a_i$, we get a contradiction to (5.1). Analogously, one shows that $h \notin \mathcal{G}$.

Assume now that we have already constructed $0 < a_1 < b_1 < \ldots < a_{i-1} < b_{i-1}$. Then the elements h_ν for $\nu < b_{i-1}$ are already determined. Consider the following finite variation

$$\tilde{f} := (h_0, h_1, \ldots, h_{b_{i-1}-1}, f_{b_{i-1}}, f_{b_{i-1}+1}, \ldots)$$

of the family f. Since $f \notin \mathcal{F}$ and $\mathcal{F}$ is closed under finite variation, we have $\tilde{f} \notin \mathcal{F}$. Therefore, $\tilde{f} \notin \mathcal{F}_{ij}$ for some j. As $\mathcal{F}_{ij}$ is a cylinder, there exists N such that $\mathcal{F}_{ij} = \pi_n^{-1}(\pi_n(\mathcal{F}_{ij}))$ for all $n \geq N$. Now choose $a_i := \max\{b_{i-1}+1, j, N\}$. Then we have

$$\pi_{a_i}^{-1}(\pi_{a_i}(\mathcal{F}_{ia_i})) \subseteq \pi_{a_i}^{-1}(\pi_{a_i}(\mathcal{F}_{ij})) = \mathcal{F}_{ij} \ .$$

Therefore, $\pi_{a_i}(\tilde{f}) \notin \pi_{a_i}(\mathcal{F}_{ia_i})$. The corresponding extension of the sequence h up to index $a_i - 1$ therefore satisfies the desired property. The index b_i can be found similarly by considering the finite variation $\tilde{g} = (h_0, \ldots, h_{a_i-1}, g_{a_i}, g_{a_i+1}, \ldots)$ of the family g. $\qquad\square$

By induction one can easily generalize Thm. 5.5 to an arbitrary finite number of σ-limit sets.

Corollary 5.6 *Let $\mathcal{F}_1, \ldots, \mathcal{F}_s$ be σ-limit sets of $\Omega^{\mathbb{N}}$ which are closed under finite variation and let $f_i \in \Omega^{\mathbb{N}} \setminus \mathcal{F}_i$ for $i = 1, \ldots, s$. Then there exists $h \in \Omega^{\mathbb{N}}$ satisfying $h \leq_p f_1 \cup \ldots \cup f_s$ and $h \notin \mathcal{F}_i$ for all i.*

Example 5.7 Consider $\Omega = \{0, 1\}$ with the natural order $\leq$. We define the support of a family $h \in \Omega^{\mathbb{N}}$ as $\{\nu \mid h_\nu \neq 0\}$. The families with finite support form a σ-limit set $\mathcal{G}$. We claim that the complement $\mathcal{F}$ of $\mathcal{G}$ is not a σ-limit set. In fact, otherwise, Thm. 5.5 with the constant families $f = (0) \notin \mathcal{F}$ and $g = (1) \notin \mathcal{G}$ would imply the existence of a family $h \notin \mathcal{F} \cup \mathcal{G}$, which is absurd. This example also shows that the σ-limit sets are not closed under the formation of countable intersections: we have $\mathcal{F} = \cap_n \mathcal{F}_n$, where $\mathcal{F}_n$ denotes the σ-limit set $\mathcal{F}_n := \cup_{m \geq n}(\Omega^m \times \{1\} \times \Omega^{\mathbb{N}})$.

5.3 An Abstract Embedding Theorem

Again let a quasi-ordered set $(\Omega, \leq)$ be fixed and denote by $\leq_p$ the corresponding abstract p-projection. We extend our discussion to any quasi-order on $\Omega^{\mathbb{N}}$ which satisfies certain compatibility conditions.

Definition 5.8 A quasi-order $\leq_c$ of $\Omega^{\mathbb{N}}$ is called *compatible*, iff the following conditions are satisfied:

(1) $\forall f, g : f \leq_p g \Rightarrow f \leq_c g$.

(2) $\forall f, g, h : f \leq_c h, g \leq_c h \Rightarrow f \cup g \leq_c h$.

(3) $\{h \mid h \leq_c g\}$ and $\{h \mid f \leq_c h\}$ are σ-limit sets for all $f, g \in \Omega^{\mathbb{N}}$.

Observe that $\leq_p$ is a compatible quasi-order by Lemma 5.4.

In the sequel, let a compatible quasi-order $\leq_c$ on $\Omega^{\mathbb{N}}$ be fixed. We call two families f and g *c-equivalent* iff $f \leq_c g$ and $g \leq_c f$. The corresponding equivalence classes are a union of certain p-degrees and called *c-degrees*. $f <_c g$ shall mean that $f \leq_c g$, but not $g \leq_c f$. We say that $f <_p g$ holds *strongly* iff $f \leq_p g$ and $f <_c g$.

Let $(X, \subseteq)$ be a poset. A map $\varphi \colon X \to \Omega^{\mathbb{N}}$ is called an *embedding* of X in $\Omega^{\mathbb{N}}$ (with respect to $\leq_c$) if $x \subseteq y$ implies $\varphi(x) \leq_c \varphi(y)$ and vice versa. We call φ a *strong embedding* iff φ is an embedding and $x \subseteq y$ implies even $\varphi(x) \leq_p \varphi(y)$.

The goal of this section is to prove the following abstract embedding theorem.

Theorem 5.9 *For any countable poset $(X, \subseteq)$ and elements $f, g \in \Omega^{\mathbb{N}}$ with $f <_c g$ there is an embedding $X \to \{h \mid f <_c h <_c g\}$. If additionally $f \leq_p g$, then there is a strong embedding $X \to \{h \mid f <_p h <_p g\}$.*

The proof will be based on a sophisticated application of our abstract diagonalization theorem 5.5. In the next lemma, we settle the special case where X consists of one point only.

Lemma 5.10 *For $f, g \in \Omega^{\mathbb{N}}$ satisfying $f <_c g$ there exists $h \in \Omega^{\mathbb{N}}$ such that $f <_c h <_c g$. If also $f \leq_p g$, then we may additionally achieve that $f \leq_p h \leq_p g$.*

Proof. $\mathcal{F} := \{h \mid g \leq_c h \cup f\}$ and $\mathcal{G} := \{h \mid h \leq_c f\}$ are σ-limit sets, as $\leq_c$ is compatible. (Observe that $\mathcal{H} := \{h' \mid g \leq_c h'\}$ and thus $\mathcal{F} = \{h \mid h \cup f \in \mathcal{H}\}$ is a σ-limit set by the remark following Lemma 5.4.) Moreover, $\mathcal{F}$ and $\mathcal{G}$ are closed under finite variation (use property (1) in Def. 5.8). By our assumption $f <_c g$ we have $f \notin \mathcal{F}$ and $g \notin \mathcal{G}$.

Theorem 5.5 implies the existence of some $h' \in \Omega^{\mathbb{N}}$ satisfying $h' \leq_p f \cup g$ and $h' \notin \mathcal{F} \cup \mathcal{G}$. Now put $h := h' \cup f$. Using property (b) in Def. 5.8 we conclude from $f \leq_c g$ that $f \leq_p h \leq_c g$. On the other hand, since $h' \notin \mathcal{F} \cup \mathcal{G}$, we have the strict inequalities $f <_c h <_c g$. Of course, if additionally $f \leq_p g$, then we even get $h \leq_p g$. $\qquad\square$

We need two further auxiliary results.

Lemma 5.11 *Any countable poset $(X, \subseteq)$ can be embedded in a countable lattice.*

Proof. For $y \in X$ denote by $X_y := \{x \in X \mid x \subseteq y\}$ the initial segment of y. Let $\mathcal{A}$ denote the Boolean subalgebra generated by all initial segments. $\mathcal{A}$ is a countable lattice with respect to inclusion and the map $X \to \mathcal{A}, y \mapsto X_y$ obviously defines an order isomorphism. $\qquad\square$

Lemma 5.12 *Let $(X, \subseteq)$ be a countable lattice. Then there exists an enumeration $x_0, x_1, x_2, \ldots$ of X such that each $X_n := \{x_0, \ldots, x_n\}$ is closed under taking meets: that is, $x \cap y \in X_n$ for all $x, y \in X_n$.*

Proof. We may assume that X is infinite. Let $\nu \colon X \to \mathbb{N}$ be any enumeration of X. We proceed recursively: set $x_0 := \nu^{-1}(0)$. Assume now that $x_0, \ldots, x_{n-1}$ are already constructed. Let $z_n \in X \setminus X_{n-1}$ such that $\nu(z_n)$ is minimal and consider the finite set

$$M_n := \{z_n \cap x \mid x \in X_{n-1},\ z_n \cap x \notin X_{n-1}\} \ .$$

If M_n is empty, we put $x_n := z_n$. Then X_n is obviously closed under taking meets.

Otherwise, let $z_n \cap x$ be a minimal element of M_n w.r.t. $\subseteq$ and define $x_n := z_n \cap x$. To show that X_n is closed under the formation of meets, let $a \in X_{n-1}$. We have $x_n \cap a = (z_n \cap x) \cap a = z_n \cap x'$, where $x' := x \cap a$ is in X_{n-1} by the induction hypothesis. By the minimality in M_n we see that in fact $x_n \cap a \in X_n$.

It remains to show that $x_0, x_1, \ldots$ exhaust all elements of X. If this were not the case, take $z \in X \setminus \{x_0, x_1, \ldots\}$ with a minimum value of ν. Hence there exists n_0 such that

$$\{y \mid \nu(y) < \nu(z)\} \subseteq \{x_0, \ldots, x_{n_0}\} \ .$$

Therefore, we have $z_n = z$ and $M_n \neq \emptyset$ for all $n > n_0$. It is easy to check that $M_{n+1} \subseteq M_n \setminus \{x_n\}$ for $n > n_0$. Thus we would obtain a infinite strictly descending chain of subsets in the finite set M_{n_0}, which is absurd. $\square$

Proof. (of Thm. 5.9) Let $(X, \subseteq)$ be a countable poset and assume that $f <_p g$ holds strongly. (The case where we only know that $f <_c g$ can be settled similarly.)

By the Lemmas 5.11 and 5.12 we may assume that $(X, \subseteq)$ is a lattice and that $x_0, x_1, \ldots$ is an enumeration of X such that each $X_n = \{x_0, \ldots, x_n\}$ is closed under the formation of meets.

By induction on n, we shall construct maps $\varphi_n \colon X_n \to \Omega^{\mathbb{N}}$ satisfying $\varphi_{n-1} = \varphi_n \mid_{X_{n-1}}$ and such that the following two properties are satisfied:

(a) $f <_p \bigcap_{x \in X_n} \varphi_n(x)$ strongly , $\bigcup_{x \in X_n} \varphi_n(x) <_p g$ strongly .

(b) For all $x, y, y_1, \ldots, y_s \in X_n$ we have

$$x \subseteq y \Rightarrow \varphi_n(x) \leq_p \varphi_n(y),\ \varphi_n(x) \leq_c \varphi_n(y_1) \cup \ldots \cup \varphi_n(y_s) \Rightarrow x \subseteq y_1 \cup \ldots \cup y_s.$$

The induction start where $n = 0$ is guaranteed by Lemma 5.10. Now let $n > 0$ and assume that $x_0, \ldots, x_{n-1}$ satisfying the claim are already constructed. To simplify notation we write $\varphi := \varphi_{n-1}$, $A := X_{n-1}$, and $z := x_n$.

Let $a_1, \ldots, a_p$ denote the maximal elements of A which are smaller than z and $b_1, \ldots, b_q$ be the minimal elements of A which are bigger than z. Thus for all $x, y \in A$ the relation $x \subseteq z$ implies $x \subseteq a_i$ for some i and $z \subseteq y$ implies $b_j \subseteq y$ for some j.

We are going to distinguish several cases.

Case 1: $p \geq 1, q \geq 1$.

We set $u := \varphi(a_1) \cup \ldots \cup \varphi(a_p)$ and $o := \varphi(b_1 \cap \ldots \cap b_q)$. (Note that this is well defined, as A is closed under taking meets!) We have $u \leq_p o$. For $x \in A$ and $y = (y_1, \ldots, y_s) \in A^s$ we define the set

$$\mathcal{G}_y := \{h \mid h \leq_c \varphi(y_1) \cup \ldots \cup \varphi(y_s)\}$$

if $z \not\subseteq y_1 \cup \ldots \cup y_s$, $s \geq 1$, and we define

$$\mathcal{F}_{x,y} := \{h \mid \varphi(x) \leq_c h \cup u \cup \varphi(y_1) \cup \ldots \cup \varphi(y_s)\}$$

if $x \not\subseteq z \cup y_1 \cup \ldots \cup y_s$, $s \geq 0$. All these sets $\mathcal{G}_y, \mathcal{F}_{x,y}$ are σ-limits and closed under finite variation. There is at most a finite number of them.

We claim that u lies in none of the sets $\mathcal{F}_{x,y}$. Otherwise, we would have $\varphi(x) \leq_c u \cup \varphi(y_1) \cup \ldots \cup \varphi(y_s)$ which implied by (b) that $x \subseteq a_1 \cup \ldots \cup a_p \cup y_1 \cup \ldots \cup y_s \subseteq z \cup y_1 \cup \ldots \cup y_s$, contradicting our assumption. In the same way one sees that o lies in none of the sets $\mathcal{G}_y$.

By Cor. 5.6 of the abstract diagonalization theorem, there is some h which lies in none of the sets $\mathcal{G}_y$ and $\mathcal{F}_{x,y}$ and such that $h \leq_p u \cup o \leq_p o$. We extend now the map $\varphi = \varphi_{n-1}$ to X_n by setting $\varphi_n(z) := h \cup u$. Then condition (a) of the inductive claim is obviously satisfied. Moreover, we have

$$\varphi(a_i) \leq_p u \leq_p \varphi_n(z) \leq_p o \leq \varphi(b_j)$$

for all i, j, which, together with the inductive hypothesis, shows that $x \subseteq y$ implies $\varphi_n(x) \leq_p \varphi_n(y)$ for all $x, y \in X_n$. To prove the second part of the claim (b) assume that $\varphi_n(z) \leq_c \varphi(y_1) \cup \ldots \cup \varphi(y_s)$. If we had $z \not\subseteq y_1 \cup \ldots \cup y_s$, then we would obtain the contradiction $h \in \mathcal{G}_y$. Similarly, $\varphi(x) \leq_c \varphi_n(z) \cup \varphi(y_1) \cup \ldots \cup \varphi(y_s)$ implies $x \subseteq z \cup y_1 \cup \ldots \cup y_s$, since $h \notin \mathcal{F}_{x,y}$. This proves part (b) of the claim.

Case 2: $p \geq 1, q = 0$.

By Lemma 5.10 there exists g' such that $\cup_{x \in A} \varphi(x) <_p g' <_p g$ holds strongly. We set $u := \varphi(a_1) \cup \ldots \cup \varphi(a_p)$ but now we define $o := g'$. For the sets $\mathcal{G}_y$ and $\mathcal{F}_{x,y}$ introduced as above we have $u \notin \mathcal{F}_{x,y}$ and $o \notin \mathcal{G}_y$. By Cor. 5.6 there is some h lying in none of the $\mathcal{G}_y, \mathcal{F}_{x,y}$ and such that $h \leq_p u \cup o \leq_p o$, and we define $\varphi_n(z) := h \cup u$. Then condition (a) of the claim remains valid, as $\cup_{x \in X_n} \leq_p g'$ and $g' <_p g$ strongly holds. The remaining conditions can be checked as in Case 1.

Case 3: $p = 0, q \geq 1$.

By Lemma 5.10 there exists f' such that $f <_p f' <_p \cap_{x \in A}$ holds strongly. We take $u := f'$ and $o := \varphi(b_1 \cap \ldots \cap b_q)$. The sets $\mathcal{G}_y$ and $\mathcal{F}_{x,y}$ are defined as before, but using the element $u = f'$. We proceed now similarly as before.

Case 4: $p = 0, q = 0$.

By Lemma 5.10 there exist f', g' satisfying $f <_p f' <_p \cap_{x \in A}$ and $\cup_{x \in A} \varphi(x) <_p g' <_p g$ strongly. Take $u := f'$, $o := g'$, and proceed similarly as before. $\qquad\square$

5.4 Structure of Valiant's Complexity Classes

In this section, we apply our previous results to the setting of Valiant. Let $\Omega := k[X_1, X_2, \ldots]$ denote the polynomial ring over a fixed field k in countably many variables X_i and consider the projection $\leq$, which is a quasi-order on Ω. (Recall that $f \leq g$ iff f can be obtained from g by a substitution of its variables by variables or constants in k.) The corresponding quasi-order $\leq_p$ on $\Omega^{\mathbb{N}}$ is the usual p-projection.

To avoid confusions, we remark that in the future symbols like $f, g, h, \ldots$ will be used to denote either polynomials or sequences of polynomials; it will always be clear from the context what is meant.

We introduce the concept of oracle computations. Let a polynomial g in $k[X_1, \ldots, X_s]$ be given. We consider straight-line programs which, beside the usual arithmetic operations, have the ability to evaluate the "oracle polynomial" g at previously computed values at unit cost. This can easily be formalized by considering straight-line programs Γ of type $\{+, -, *, o\}$, where the symbol o stands for the oracle operation of arity s.

Definition 5.13 The *oracle complexity* $L^g(f_1, \ldots, f_t)$ of a set of polynomials $f_1, \ldots, f_t \in \Omega$ with respect to the oracle polynomial g is the minimum number of arithmetic operations $+, -, *$ and evaluations of g (at previously computed values) that are sufficient to compute the f_j from the indeterminates X_i and constants in k.

We introduce next the notion of *c*-reduction, which can be seen as an analogue of the polynomial Turing reduction for Valiant's setting. (*c* is an acronym for computation.) One might also interpret the p-projection as an analogue of the polynomial many-one reduction, however, the p-projection is much finer.

Definition 5.14 Let $f = (f_n)$, $g = (g_n) \in \Omega^{\mathbb{N}}$. We call f a *c-reduction* (or polynomial oracle reduction) of g, shortly $f \leq_c g$, iff there is a p-bounded function $t \colon \mathbb{N} \to \mathbb{N}$ such that the map $n \mapsto L^{g_{t(n)}}(f_n)$ is p-bounded.

It easy to check that $\leq_c$ is a quasi-order of $\Omega^{\mathbb{N}}$. Note that for a p-family f we have $f \leq_c 0$ iff f is p-computable.

Lemma 5.15 *The c-reduction $\leq_c$ is a compatible quasi-order on $\Omega^{\mathbb{N}}$.*

Proof. The verification of conditions (1) and (2) of Def. 5.8 is straightforward. Condition (c) will be shown similarly as in the proof of Lemma 5.4. We can express $h \leq_c g$ by the following predicate

$$\exists c \; \forall n \; \exists m : m \leq n^c + c \; \wedge \; L^{g_m}(h_n) \leq n^c + c \; .$$

If we write

$$V(c,n) := \{u \in \Omega \mid \exists m \leq n^c + c : L^{g_m}(u) \leq n^c + c\} \; ,$$

then we have $\{h \mid h \leq_c g\} = \bigcup_c \prod_n V(c,n)$, which shows that this is a σ-limit set.

On the other hand, we may write $\{h \mid f \leq_c h\}$ as the countable union over all c of the following limits of cylinders

$$\bigcap_n \bigcup_{m \leq n^c + c} \left(\Omega^m \times \{v \in \Omega \mid L^v(f_n) \leq n^c + c\} \times \Omega^{\mathbb{N}} \right) \; .$$

$\square$

Corollary 5.16 *The set of p-families as well as the classes* VP *and* VNP *are σ-limit sets.*

Proof. We leave it to the reader to check that the set $\mathcal{P}$ of p-families is a σ-limit. Let g be VNP-complete w.r.t. p-projection. We have $VP = \{f \in \Omega^{\mathbb{N}} \mid f \leq_c 0\} \cap \mathcal{P}$ and $VNP = \{f \in \Omega^{\mathbb{N}} \mid f \leq_p g\}$. Thus we may conclude from Lemma 5.15 and the fact that the p-projection is compatible, that both of these sets are σ-limits. $\square$

Let us call a p-degree or a c-degree *p-definable* iff it contains a p-definable family. Note that a p-definable p-degree consists of p-definable families only, whereas a p-definable c-degree might also contain families which are not in VNP. This is because $f \leq_c g$ and $g \in VNP$ might not imply that $f \in VNP$. We denote by $\mathcal{PD}_p$ the set of p-degrees of p-definable families and by $\mathcal{PD}_c$ the set of c-degrees of p-definable families.

Remark 5.17   (1) The poset $\mathcal{PD}_p$ has a unique maximal p-degree which consists of the VNP-complete families with respect to p-projection. Any family (f_n) of constants (i.e., $f_n \in k$ for all n) constitutes a minimal p-degree in $\mathcal{PD}_p$, and these are all the minimal p-degrees in $\mathcal{PD}_p$. (Hence $\mathcal{PD}_p$ has at least the cardinality of the continuum.)

(2) The poset $\mathcal{PD}_c$ has a unique maximal c-degree which consists of the VNP-complete families with respect to c-reduction. The complexity class VP forms the unique minimal c-degree in $\mathcal{PD}_c$. Valiant's hypothesis "VNP $\neq$ VP" means that $\mathcal{PD}_c$ consists of more than one element.

The main result of this section is analogous to that of Ladner's work [70]. It follows now easily from our abstract embedding theorem 5.9.

Theorem 5.18 *Any countable poset can be embedded in $\mathcal{PD}_p$. If Valiant's hypothesis is true, then any countable poset can be embedded in $\mathcal{PD}_c$.*

Note that the result for $\mathcal{PD}_p$ is unconditional due to Remark 5.17(1).

Corollary 5.19 *If Valiant's hypothesis is true, then there is a p-definable family which is neither p-computable nor VNP-complete with respect to c-reduction.*

We finally show that an analogue of Schöning's general minimal pair theorem [97] holds in Valiant's setting. We call a pair of families $\varphi, \psi \in \Omega^{\mathbb{N}}$ a *minimal pair* for VP iff φ and ψ are not contained in VP and

$$\forall h \in \Omega^{\mathbb{N}} : h \leq_c \varphi \ \wedge \ h \leq_c \psi \Longrightarrow h \in \text{VP} \ .$$

Theorem 5.20 *Assume that $\mathcal{F} \subseteq \Omega^{\mathbb{N}}$ is a σ-limit set containing VP which is closed under finite variation, and let $f, g \in \Omega^{\mathbb{N}} \setminus \mathcal{F}$. Then there exist $\varphi, \psi \in \Omega^{\mathbb{N}} \setminus \mathcal{F}$ such that $\varphi \leq_p f$, $\psi \leq_p g$, and such that φ, ψ is a minimal pair for VP.*

Proof. Let $\mathcal{F} = \bigcup_i \bigcap_j \mathcal{F}_{ij}$ with cylinders $\mathcal{F}_{ij}$ satisfying $\mathcal{F}_{ij} \supseteq \mathcal{F}_{i\,j+1}$. By induction, we will construct a sequence $0 = a_{10} < a_{11} < a_{12} < a_{13} < a_{14} < a_{15} < a_{20} < \ldots < a_{25} < \ldots$ of natural numbers satisfying the requirements below. We define families φ and ψ corresponding to the sequence $(a_{ij})_{1 \leq i, 0 \leq j \leq 5}$ by setting

$$\varphi_\nu := \left\{ \begin{array}{ll} f_\nu & \text{if } \exists i : a_{i0} \leq \nu < a_{i1} \\ 0 & \text{otherwise} \end{array} \right. \qquad \psi_\nu := \left\{ \begin{array}{ll} g_\nu & \text{if } \exists i : a_{i3} \leq \nu < a_{i4} \\ 0 & \text{otherwise.} \end{array} \right.$$

The requirements are:

$$(0) \quad \pi_{a_{i1}}(\varphi) \notin \pi_{a_{i1}}(\mathcal{F}_{ia_{i1}}) \qquad\qquad (3) \quad \pi_{a_{i4}}(\psi) \notin \pi_{a_{i4}}(\mathcal{F}_{ia_{i4}})$$

$$(1) \quad \max_{m \leq a_{i1}} L(f_m) \leq a_{i2} \qquad\qquad\quad (4) \quad \max_{m \leq a_{i4}} L(g_m) \leq a_{i5}$$

$$(2) \quad 2^{a_{i2}} \leq a_{i3} \qquad\qquad\qquad\qquad\quad (5) \quad 2^{a_{i5}} \leq a_{i+1\,0} \ .$$

As in the proof of the abstract diagonalization theorem 5.5, one can show that it is possible to construct a sequence (a_{ij}) satisfying all these requirements. (Only conditions (0) and (3) require some attention.)

Let us show that φ, ψ have the desired properties. It is clear that $\varphi \leq_p f$ and $\psi \leq_p g$. Moreover, we have $\varphi, \psi \notin \mathcal{F}$ due to conditions (0) and (3). It remains to prove that φ, ψ is a minimal pair. So let us assume that $h \leq_c \varphi$ and $h \leq_c \psi$ for some $h \in \Omega^{\mathbb{N}}$. Then there exist p-bounded functions $u, v, w \colon \mathbb{N} \to \mathbb{N}$ satisfying

$$L^{\varphi_{u(n)}}(h_n) \leq w(n) \ , \quad L^{\psi_{v(n)}}(h_n) \leq w(n) \ .$$

It suffices to verify that $L(h_n) \leq n w(n)$ for sufficiently large n.

We are going to distinguish two cases. Suppose first that $a_{i2} \leq n < a_{i5}$. We may assume that $a_{j0} \leq u(n) < a_{j1}$ for some j, since otherwise $\varphi_{u(n)} = 0$ and we are done. Thus $\varphi_{u(n)} = f_{u(n)}$. For sufficiently large n we have by condition (5) that $u(n) \leq 2^n < 2^{a_{i5}} \leq a_{i+1\,0}$. This implies that $j \leq i$, hence $u(n) < a_{i1}$. Therefore, using condition (1), we have $L(f_{u(n)}) \leq a_{i2} \leq n$. We conclude that indeed

$$L(h_n) \leq L^{\varphi_{u(n)}}(h_n)\, L(\varphi_{u(n)}) \leq nw(n) \ .$$

The discussion of the other case where $a_{i5} \leq n < a_{i+1\,2}$ is similar and left to the reader. $\qquad\square$

By applying the theorem to $\mathcal{F} = \mathrm{VP}$ and choosing $f = g$ to be VNP-complete we obtain the following corollary. (Note that VP is a σ-limit set by Corollary 5.16.)

Corollary 5.21 *There exists a minimal pair* φ, ψ *for* VP *in* VNP, *provided* $\mathrm{VP} \neq \mathrm{VNP}$.

Problem 5.1 Is the class VNP closed under c-reduction?

5.5   A Specific Family Neither Complete Nor p-Computable

For $1 \leq i < j \leq n$ let X_{ij} be independent indeterminates and set $X_{ji} := X_{ij}$. Moreover, let q be a power of the prime p. The *cut enumerator* Cut_n^q is the following multivariate polynomial over the finite field $\mathbb{F}_q$

$$\mathrm{Cut}_n^q := \sum_S \prod_{i \in A, j \in B} X_{ij}^{q-1} \ ,$$

where the sum is over all cuts $S = \{A, B\}$ of the complete graph K_n on the set of nodes $\underline{n} := \{1, 2, \ldots, n\}$. (A cut of a graph is a partition of its set of nodes into two nonempty subsets.) It is easy to see that $\mathrm{Cut}^q := (\mathrm{Cut}_n^q)$ is a p-definable family.

To motivate this definition, consider a complete graph $K_n = (\underline{n}, E_n)$ endowed with a weight function $w \colon E_n \to \mathbb{N}$. We define the weight $w(S)$ of a cut $S = \{A, B\}$ as the sum of the weights of all edges separated by S. Let $c(s)$ denote the number of cuts of weight s. (Notice that the w_{ij} are interpreted here as additive weights, whereas the X_{ij} above are viewed as multiplicative weights.) Under the substitution $X_{ij} \mapsto x_{ij} := T^{w_{ij}}$, T being a formal variable, the cut polynomial Cut_n^q becomes

$$(5.2) \qquad \mathrm{Cut}_n^q(x) = \sum_S T^{(q-1)w(S)} = \sum_s (c(s) \bmod p)\, T^{(q-1)s} \ ,$$

which is essentially the generating function of the sequence $(c(s) \bmod p)_s$.

The main result of this section states that Cut^q is an explicit example of a p-family, which is neither p-computable nor complete in VNP. For the definition of the complexity classes $\mathrm{Mod}_p\mathrm{NP}/\mathrm{poly}$ see below.

Theorem 5.22 *The family of cut enumerators Cut^q over a finite field $\mathbb{F}_q$ is neither p-computable nor VNP-complete with respect to c-reduction, provided $\mathrm{Mod}_p\mathrm{NP} \not\subseteq \mathrm{P}/\mathrm{poly}$. The latter condition is satisfied if the polynomial hierarchy does not collapse at the second level.*

The proof of this theorem requires two auxiliary results. The first of them states that the cut polynomial $\mathrm{Cut}_n^q(x)$ can be evaluated over $\mathbb{F}_q$ by Boolean circuits of polynomial size. The reader should be aware that this does not necessarily imply that the cut polynomial can also be evaluated by arithmetic circuits of p-bounded size (i.e., the p-computability of the family Cut^q).

Lemma 5.23 *To a symmetric matrix $x \in \mathbb{F}_q^{n \times n}$ we assign the graph $G(x)$ on the set of nodes $\underline{n}$ by requiring that $\{i, j\}$ is an edge iff $x_{ij} = 0$. Then we have*

$$\mathrm{Cut}_n^q(x) = 2^{N(x)-1} - 1 \bmod p \ ,$$

where $N(x)$ is the number of connected components of $G(x)$. In particular, the value $\mathrm{Cut}_n^q(x)$ can be computed from a symmetric $x \in \mathbb{F}_q^{n \times n}$ in polynomial time by a Turing machine.

Proof. For any nonzero $\lambda \in \mathbb{F}_q$ we have $\lambda^{q-1} = 1$ by Fermat's theorem. Therefore, a partition $\{A, B\}$ of $\underline{n}$ contributes to $\mathrm{Cut}_n^q(x)$ either zero or one. The contribution is one iff $x_{ij} \neq 0$ for all $i \in A$, $j \in B$, which is the case iff none of the nodes of A is connected with any node in B in the graph $G(x)$. This in turn means that A and B are both a union of certain connected components of the graph $G(x)$. The number of such partitions clearly equals $2^{N(x)-1} - 1$, where $N(x)$ is the number of connected components of $G(x)$. This proves the lemma. $\qquad\square$

It is now time to recall a few facts from discrete complexity theory (compare Sect. 4.3). For a prime number p the class $\mathrm{Mod}_p\mathrm{NP}$ is defined as the set of languages $\{x \in \{0,1\}^* \mid \phi(x) \equiv 1 \bmod p\}$, where $\phi \colon \{0,1\}^* \to \mathbb{N}$ is a function in #P. We remark that if $\phi \colon \{0,1\}^* \to \mathbb{N}$ is #P-complete with respect to parsimonious reductions, then the corresponding language $\{x \mid \phi(x) \equiv 1 \bmod p\}$ is $\mathrm{Mod}_p\mathrm{NP}$-complete (with respect to polynomial many-one reductions). We denote by $\mathcal{C}/\mathrm{poly}$ the nonuniform version of the complexity class $\mathcal{C}$.

The counting problem #Cut is the following: given a complete graph K_n with a weight function $w \colon E_n \to \mathbb{N}$ and $s \in \mathbb{N}$, what is the number of cuts of weight s? Hereby, we assume the edge weights to be encoded in unary. The related decision problem $\mathrm{Mod}_p\mathrm{Cut}$ just asks whether the number of cuts of weight s is congruent to 1 modulo p. This problem is clearly in the class $\mathrm{Mod}_p\mathrm{NP}$.

It is well known that the computation of a cut of maximal weight of a given graph is NP-hard. By a straightforward modification of the proof of this fact given in Papadimitriou [87, p. 191], one can strengthen this as follows. We will provide the detailed proof of this claim at the end of this section.

Lemma 5.24 $\#$CUT *is* $\#$P*-complete with respect to parsimonious reductions. Thus* Mod_pCUT *is* Mod_pNP*-complete.*

Proof. (of Thm. 5.22) Let L be a language in Mod_pNP, say $L = \{x \in \{0,1\}^* \mid \phi(x) \equiv 1 \bmod p\}$, where $\phi\colon \{0,1\}^* \to \mathbb{N}$ is in the class $\#$P. By Thm. 4.5(2), there exists a p-definable family (f_n) over $\mathbb{F}_p$ such that $f_n \in \mathbb{F}_p[X_1, \ldots, X_n]$ and

$$\forall n \; \forall x \in \{0,1\}^n : f_n(x) = \phi(x) \bmod p \ .$$

Assume now that Cut^q is VNP-complete over $\mathbb{F}_q$ with respect to c-reduction. Then we have $(f_n) \leq_c \mathrm{Cut}^q$, hence there is a p-bounded function $t\colon \mathbb{N} \to \mathbb{N}$ such that $L^{\mathrm{Cut}^q_{t(n)}}(f_n)$ is p-bounded. Lemma 5.23 tells us that $\mathrm{Cut}^q_{t(n)}$ can be evaluated over $\mathbb{F}_q$ by Boolean circuits of p-bounded size in n. Hence, by simulating straight-line programs by Boolean circuits, we can design for each n a Boolean circuit C_n of p-bounded size in n, which computes $f_n(x)$ from $x \in \mathbb{F}_q^n$. This implies that the language L is contained in P/poly. We therefore arrive at the conclusion Mod_pNP $\subseteq$ P/poly.

For fixed $m, n \geq 1$ consider a field extension $K = \mathbb{F}_q(\xi)$ of $\mathbb{F}_q$ of degree $(q-1)m\binom{n}{2}$. To an instance $w\colon E_n \to \mathbb{N}$ of $\#$CUT satisfying $\max w \leq m$ we assign the symmetric matrix $x \in K^{n \times n}$ defined by $x_{ij} := \xi^{w_{ij}}$. Then we have by (5.2)

$$\mathrm{Cut}^q_n(x) = \sum_s (c(s) \bmod p)\, \xi^{(q-1)s} \ ,$$

where $c(s)$ is the number of cuts in K_n of weight s. The coefficients $c(s) \bmod p$ are uniquely determined by $\mathrm{Cut}^q_n(x)$ since the above summation is over all $s < m\binom{n}{2}$.

Assume now that Cut^q is p-computable over $\mathbb{F}_q$. Hence for each n there is a straight-line program Γ_n of p-bounded size in n, which computes $\mathrm{Cut}^q_n(X)$ from constants in $\mathbb{F}_q$ and the indeterminates X_{ij} in the polynomial ring $\mathbb{F}_q[X_{ij} \mid 1 \leq i, j \leq n]$. By the universal property of the polynomial ring, Γ_n will compute $\mathrm{Cut}^q_n(x)$ in the $\mathbb{F}_q$-algebra K from the same constants and $x \in K^{n \times n}$. We may simulate this computation by a Boolean circuit of p-bounded size, since the arithmetic operations in K can be simulated by p-bounded circuits. Here it is important to note that the degree of the field extension $K/\mathbb{F}_q$ is p-bounded, as m is assumed to be encoded in unary (see the definition of $\#$CUT). In this way, we could solve the Mod_pCUT problem in nonuniform polynomial time. As Mod_pCUT is Mod_pNP-complete by Lemma 5.24, this would imply that Mod_pNP $\subseteq$ P/poly.

It remains to show that $\mathrm{Mod}_p\mathrm{NP} \subseteq \mathrm{P}/\mathrm{poly}$ implies the collapse of the polynomial hierarchy at the second level. From Thm. 4.10 we know that

$$\mathrm{NP}/\mathrm{poly} \subseteq \mathrm{Mod}_p\mathrm{NP}/\mathrm{poly} \ .$$

Therefore, $\mathrm{Mod}_p\mathrm{NP} \subseteq \mathrm{P}/\mathrm{poly}$ implies the the inclusion $\mathrm{NP}/\mathrm{poly} \subseteq \mathrm{P}/\mathrm{poly}$. By a well-known result of Karp and Lipton [61], this would have the collapse of the polynomial hierarchy at the second level as a consequence.   □

We remark that one can prove the absolute statement that Cut^q is not VNP-complete with respect to p-projection.

Problem 5.2 Is Cut^2, interpreted as family over the rationals, VNP-complete?

Problem 5.3 Find other examples of specific p-definable families, which are not complete.

We supply now the proof of Lemma 5.24. Let #Sat denote the problem to count the number of satisfying assignments for a given Boolean formula in conjunctive normal form. The problem #Sat is known to be #P-complete with respect to parsimonious reductions. Consider now the auxiliary counting problem #Naesat which is defined as follows: given a set of Boolean variables and a set of clauses each consisting of exactly three literals, compute N, where $2N$ equals the number of truth assignments of the variables such that in none of the clauses all three literals have the same truth value. (Note that the latter number must always be even!) The corresponding decision problem is denoted by Naesat (not-all-equal Sat).

Lemma 5.25 *There is a parsimonious reduction from* #Sat *to* #Naesat.

Proof. The reduction from Circuit Sat to Naesat given in Example 8.3 (p. 163) and Thm. 9.3 (p. 187) of Papadimitriou [87] is easily checked to be parsimonious. On the other hand, Sat can be parsimoniously reduced to Circuit Sat in an obvious way.   □

To prove Lemma 5.24 it suffices now to show the next lemma.

Lemma 5.26 *There is a parsimonious reduction from* #Naesat *to* #Cut.

Proof. We slightly modify the reduction from Naesat to Max Cut from Papadimitriou [87, Thm. 9.5, p. 191] in order to make it parsimonious.

Let be given a set of variables $x_1, \ldots, x_n$ and a set of clauses $C_1, \ldots, C_m$ each consisting of exactly three literals. We may assume that in no clause all literals are equal since otherwise the formula is not satisfiable in the sense of Naesat. Moreover, we may remove the clauses which contain a variable and its negation since these are always satisfiable in the sense of Naesat. Let m_3 denote the number of clauses with three different literals and m_2 be the number of clauses in which two literals coincide. We have $m = m_2 + m_3$.

Let G be the complete graph having as nodes the variables x_i and its negations $\neg x_i$. We define the weight function of G as follows. The *horizontal edges* $\{x_i, \neg x_i\}$ have the weight $m + 1$. The remaining edges edges $e = \{u, v\}$ (the *nonhorizontal* ones) have as weight the number of clauses C_j in which both of the literals u and v appear. If we express this event by $e \subseteq C_j$, we may write for such nonhorizontal edges e

$$w(e) = |\{C_j \mid e \subseteq C_j\}| \ .$$

Finally, we put $s := (m + 1)n + m_2 + 2m_3$. Note that the weight of each edge is at most $m + 1$. Thus we may encode the edge weights in unary.

Let S be a cut of G and denote by $\mathcal{E}_h$ the set of horizontal edges separated by S, and by $\mathcal{E}$ the set of nonhorizontal edges separated by S. We have

$$
\begin{aligned}
w(S) \ &= \ \sum_{e \in \mathcal{E}_h \cup \mathcal{E}} w(e) = (m + 1)|\mathcal{E}_h| + \sum_{e \in \mathcal{E}} |\{(e, C_j) \mid e \subseteq C_j\}| \\
&= \ (m + 1)|\mathcal{E}_h| + \sum_{j=1}^{m} |\{(e, C_j) \mid e \in \mathcal{E}, e \subseteq C_j\}| \\
&\leq \ (m + 1)n + (m_2 + 2m_3) = s \ .
\end{aligned}
$$

Equality holds if and only if S separates all x_i from $\neg x_i$ and if S separates the literals of any clause. This is exactly the case if S defines a truth assignment in the sense of NAESAT. (The cut S separates the true literals from the false ones.) This proves that the number of satisfying truth assignments is exactly twice the number of cuts of weight s in G. $\qquad \square$

5.6   Relativized Complexity Classes

Our investigations here are inspired by the well-known results of Baker, Gill, and Solovay [3] on relativations of the classical P-NP question.

Relativized versions of the complexity classes VP and VNP can be defined as follows.

Definition 5.27 Let h be a p-family. VP^h consists of all p-families f such that $f \leq_c h$. VNP^h is the set of all p-families $f = (f_n)$ which can be obtained from some $g = (g_n) \in \mathrm{VP}^h$ as in Def. 2.5. We call the families in VP^h and VNP^h p-computable and p-definable relative to h, respectively.

Note that VP^h and VNP^h specialize to VP and VNP, respectively, if h is p-computable. It is clear that VP^h is closed under c-reduction and VNP^h is closed under p-projection. In passing, we mention that Cor. 2.22 establishes a nontrivial property of the classes VP^h: they are closed under taking factors.

Our first goal is to establish the existence of complete families for the complexity classes VP^h and VNP^h. In particular, this gives a proof for the existence of VNP-complete families, which is independent of Valiant's intricate completeness proof for the permanent. The idea is to use a generalization of the concept of generic computations (cf. [21, Chap. 9]). In order to avoid an exponential growth of degrees, we combine this with an auxiliary result on the computation of homogeneous components (Prop. 5.28), which works by evaluation and interpolation, and requires that k contains sufficiently many points. In the sequel, we will therefore assume that k is an infinite field.

It is useful to introduce the following auxiliary notion. Let $h, f_1, \ldots, f_t$ be polynomials over the field k. We define the h-*complexity* $\mathcal{L}^h(f_1, \ldots, f_t)$ as the minimum number of multiplications and evaluations of h that are sufficient to compute all f_i from the indeterminates and constants in k (we do not allow divisions). Note that for $h = X_1 X_2$ this specializes to the multiplicative (or nonscalar) complexity. We further remark that if h is a projection of h', then we have $L^{h'} \leq L^h$ as well as $\mathcal{L}^{h'} \leq \mathcal{L}^h$.

The h-complexity may be characterized in a way similar to the multiplicative complexity. Let us define an h-*computation sequence* of length r on $X_1, \ldots, X_n$ as a sequence of polynomials $g_{-n}, g_{-n+1}, \ldots, g_r$ such that $g_{-n} = 1, g_{-n+1} = X_1, \ldots, g_0 = X_n$, and such that we have

$$(5.3) \qquad g_\rho = h\left(\sum_{j=-n}^{\rho-1} \alpha_{\rho 1 j} g_j, \ldots, \sum_{j=-n}^{\rho-1} \alpha_{\rho s j} g_j \right)$$

for all $1 \leq \rho \leq r$ and some $\alpha_{\rho\sigma j}$ in k. We say that such a sequence *computes* $f_1, \ldots, f_t$ iff all f_i are contained in the k-linear hull of $g_{-n}, \ldots, g_r$.

In what follows, we will assume that $X_1 X_2$ is a projection of h, in which case we say that h *contains the multiplication*. Then it is not hard to see that the h-complexity r of $f_1, \ldots, f_t$ equals the minimum length of an h-computation sequence which computes all f_i. Moreover, the complexity L^h and the h-complexity r are polynomially related as follows: we have $r \leq L^h(f_1, \ldots, f_t) \leq 2s(n+1)(r+1) + sr^2$, when s is the number of variables of h.

Proposition 5.28 *Let f be a polynomial in $a_1, \ldots, a_m$ and $X_1, \ldots, X_n$ having degree at most $d \geq 1$ in the X-variables. We denote by $f^{(\delta)}$ the homogeneous part of f of degree δ with respect to the X-variables. Then we have*

$$\mathcal{L}^h(\{f^{(\delta)} \mid \delta \leq d\}) \leq (1 + d \deg h)\, \mathcal{L}^h(f) \ .$$

Proof. We will use the abbreviation $f^{\leq d} := \sum_{\delta \leq d} f^{(\delta)}$ and write $D := \deg h$. Let $(g_\rho)_{\rho \geq -n}$ be an h-computation sequence of length $r := \mathcal{L}^h(f)$ as in (5.3) which computes f. We define a related sequence $(u_\rho)_{\rho \geq -n}$ by setting $u_\rho := g_\rho$ for $-n \leq \rho \leq 0$, and for $\rho > 0$

$$u_\rho = h(v_{\rho 1}, \ldots, v_{\rho s})^{\leq d} \ ,$$

where $v_{\rho\sigma} = \sum_{j=-n}^{\rho-1} \alpha_{\rho\sigma j} u_j$. It is easy to check that $u_\rho = g_\rho^{\leq d}$ for all ρ.

The homogeneous parts of f (w.r.t. X) are a k-linear combination of the homogeneous parts of the g_ρ. Therefore, it suffices to prove that all homogeneous parts of u_ρ up to degree d can be computed from the homogeneous parts of $u_{-n},\dots,u_{\rho-1}$ up to degree d by some k-linear operations and $1+dD$ evaluations of h.

The polynomial $w_\rho := h(v_{\rho 1},\dots,v_{\rho s})$ has degree at most dD. By definition, $u_\rho^{(\delta)} = w_\rho^{(\delta)}$ for $\delta \le d$. We have for $\lambda \in k$ that

$$\sum_{\delta \le dD} \lambda^\delta w_\rho^{(\delta)} = w_\rho(\lambda X) = h\big(\sum_{\delta \le d} \lambda^\delta v_{\rho 1}^{(\delta)},\dots,\sum_{\delta \le d} \lambda^\delta v_{\rho s}^{(\delta)}\big) \ .$$

Hence we can compute $w_\rho(\lambda X)$ from the $v_{\rho\sigma}^{(\delta)}$ and thus from the $u_j^{(\delta)}$ for $j < \rho$, $\delta \le d$ by k-linear operations and just one evaluation of h. We can thus compute the homogeneous parts of w_ρ as a k-linear combination of $w_\rho(\lambda X)$ for $1+dD$ different values of $\lambda \in k$ (interpolation). $\qquad\square$

In the sequel, we will use the abbreviations $X^\mu := X_1^{\mu_1}\cdots X_n^{\mu_n}$ and $|\mu| := \sum \mu_i$ for $\mu \in \mathbb{N}^n$. Moreover, we set $\deg 0 := -\infty$ for the zero polynomial.

Definition 5.29 Let a polynomial $h \in k[X_1,\dots,X_s]$ of degree D be given.

(1) We define the *generic h-computation* $(G_\rho)_{\rho \ge -n}$ on $X_1,\dots,X_n$ over k recursively as follows: $G_{-n} := 1$, $G_{-n+1} := X_1,\dots,G_0 := X_n$, and for all $\rho > 0$ we set

$$\begin{aligned} G_\rho \ :=\ &h\big(\textstyle\sum_{j=-n}^{\rho-1} a_{\rho 1 j}G_j,\dots,\sum_{j=-n}^{\rho-1} a_{\rho s j}G_j\big) \\ &+ b_\rho - h\big(\textstyle\sum_{j=-n}^{\rho-1} a_{\rho 1 j}G_{j0},\dots,\sum_{j=-n}^{\rho-1} a_{\rho s j}G_{j0}\big) \ . \end{aligned}$$

Here the $a_{\rho\sigma j}$ and b_ρ denote different indeterminates and G_{j0} is the constant term of G_j with respect to the X-variables. We shall write $G_\rho = \sum_\mu G_{\rho\mu}X^\mu$, where $G_{\rho\mu}$ depends only on the a and b-variables.

(2) The nth *generic polynomial computed relative to h* is defined as

$$C_n(h) := \sum_{|\mu| \le n}\ \sum_{\rho=-n}^{n} c_\rho G_{\rho\mu}X^\mu \ .$$

Here the c_ρ denote additional indeterminates. Thus $C_n(h)$ is the sum of the X-homogeneous parts up to degree n of $\sum_{\rho=-n}^{n} c_\rho G_\rho$.

(3) The nth *generic polynomial defined relative to h* is

$$D_n(h) := \sum_{v=0}^{n} d_v \sum_{e\in\{0,1\}^{n-v}} C_n(h)(a,b,c,X_1,\dots,X_v,e_{v+1},\dots,e_n) \ ,$$

where the d_v denote additional indeterminates. (Note that the variables $X_{v+1},\dots,X_n$ have been substituted by 0 or 1.)

The following technical lemma summarizes some of the properties of h-generic computations as well as of the polynomials $C_n(h)$ and $D_n(h)$. Recall that $h \leq h'$ means that h is a projection of h'.

Lemma 5.30 *We have for $\rho > 0$ and $\mu \neq 0$:*

(1) $G_{\rho 0} = b_\rho$.

(2) $G_{\rho \mu}$ *depends on at most* $s\rho(n + 1 + \frac{\rho - 1}{2}) + \rho$ *variables.*

(3) $\mathcal{L}^h(G_1, \ldots, G_\rho) \leq s\rho(n + \rho - 1) + 2\rho$.

(4) $\deg G_{\rho \mu} \leq 1 + 2D\rho|\mu|$.

(5) $C_n(h)$ *and* $D_n(h)$ *are polynomials in at most* $2sn^3 + 5n + 2$ *variables and have degree at most* $2Dn^2 + n + 3$.

(6) $\mathcal{L}^h(C_n(h)) \leq (1 + Dn)(2sn^2 + 4n)$.

(7) *If we abbreviate the a, b, c-variables occuring in $C_n(h)$ by $Z_1, \ldots, Z_w$, we have*

$$\{f \in k[X_1, \ldots, X_n] \mid \deg f \leq n, \ \mathcal{L}^h(f) \leq n\} \subseteq \{C_n(h)(z, X) \mid z \in k^w\} \ .$$

(8) $h \leq C_n(h)$ *if* $\deg h \leq n$. *Moreover* $C_n(h) \leq D_n(h)$.

(9) *We have* $C_n(h) \leq C_{n'}(h')$ *and* $D_n(h) \leq D_{n'}(h')$ *for all* $n \leq n'$ *and* $h \leq h'$.

Proof. Claims (1), (2), and (3) follow by straightforward calculations.

We will prove Claim (4) by induction on ρ. We note first that $\deg G_{\rho \mu} \leq 0$ for $\rho \leq 0$ and all μ. Let now $\rho > 0$ be fixed and put

$$V_\sigma := \sum_\mu V_{\sigma \mu} X^\mu := \sum_{j=-n}^{\rho-1} a_{\rho \sigma j} G_j \ .$$

By the induction hypothesis we have for all $\mu \neq 0$

$$(5.4) \qquad\qquad \deg V_{\sigma \mu} \leq 2 + 2D(\rho - 1)|\mu| \ .$$

This estimate is also true for $\mu = 0$, since $\deg G_{j0} \leq 1$. What we have to do is to prove the estimate

$$\forall \mu \neq 0 : \deg H_\mu \leq 1 + 2D\rho|\mu|$$

for the polynomial $H := \sum_\mu H_\mu X^\mu := h(V_1, \ldots, V_s)$. Clearly, it is sufficient to verify this for the power products $h = X_1^{e_1} \cdots X_s^{e_s}$ of degree at most D. In this case, we have $H = V_1^{e_1} \cdots V_s^{e_s}$, and we obtain

$$H_\nu = \sum \prod_{\sigma=1}^{s} \prod_{\epsilon=1}^{e_\sigma} V_{\sigma \mu_\sigma(\epsilon)} \ ,$$

where the sum runs over all systems of maps $\mu_\sigma\colon \{1, 2, \ldots, e_\sigma\} \to \mathbb{N}^n$ with $1 \le \sigma \le s$ satisfying $\sum_\sigma \sum_\epsilon \mu_\sigma(\epsilon) = \nu$. Using (5.4) we conclude that for $\nu \ne 0$

$$
\begin{aligned}
\deg H_\nu \;&\le\; \sum_\sigma \sum_\epsilon \big(2 + 2D(\rho - 1)|\mu_\sigma(\epsilon)|\big) \\
&\le\; 2\sum_\sigma e_\sigma + 2D(\rho - 1)\sum_\sigma \sum_\epsilon |\mu_\sigma(\epsilon)| \\
&\le\; 2D + 2D(\rho - 1)|\nu| \\
&\le\; 1 + 2D\rho|\nu| \;,
\end{aligned}
$$

which proves Claim (4).

Claim (5) follows immediately from the Claims (2) and (4), whereas Claim (6) is a consequence of Claim (3) and Prop. 5.28.

To show Claim (7) assume $f \in k[X_1, \ldots, X_n]$ such that $\mathcal{L}^h(f) \le n$. There is an h-computation sequence (g_ρ) of length n which computes f, say

$$
g_\rho = h\big(\textstyle\sum_{j<\rho} \alpha_{\rho 1 j} g_j, \ldots, \sum_{j<\rho} \alpha_{\rho s j} g_j\big) \;,
$$

and $f = \sum_{\rho=-n}^n \gamma_\rho g_\rho$, with $\alpha_{\rho\sigma j}, \gamma_\rho \in k$. Thus the substitution $a_{\rho\sigma j} \mapsto \alpha_{\rho\sigma j}$, $b_\rho \mapsto g_\rho(X = 0)$ sends G_ρ to g_ρ. If we additionally substitute the c_ρ by the γ_ρ, then the polynomial $C_n(h)$ is mapped to f, provided that $\deg f \le n$.

(8) $C_n(h) \le D_n(h)$ is obtained by substituting $d_n \mapsto 1$ and $d_v \mapsto 0$ if $v < n$. To show that $h \le C_n(h)$ consider the substitution φ which maps X_1 and c_1 to 1, sends the $a_{1,\sigma,-n+1}$ to X_σ for $1 \le \sigma \le s$, maps b_1 to $h(0, \ldots, 0)$, and sends all the remaining a-variables and c-variables to 0. All other variables shall remain invariant under φ. G_1 is mapped to $h = h(X_1, \ldots, X_s)$ under φ. We have $\deg_X G_1 \le n$ as $\deg h \le n$. From this it easily follows that $C_n(h)$ is mapped to h under φ.

Before proving Claim (9) it is useful to make the following general observation. Let $A := k[a_1, \ldots, a_m, X_1, \ldots, X_n]$ and consider a substitution (k-algebra morphism) $\varphi\colon A \to A$ which fixes the X-variables and such that $\varphi(a_i) \in k \cup \{a_1, \ldots, a_m\}$. Let $f^{(\delta)}$ denote the homogeneous part of $f \in A$ with respect to the X-variables. Then we have $\varphi(f^{(\delta)}) = \varphi(f)^{(\delta)}$. If $\sigma\colon A \to A$ is another substitution which leaves the a-variables invariant and such that $\sigma(X_i) \in k \cup \{X_1, \ldots, X_n\}$, then we have $\varphi(\sigma(f)) = \sigma(\varphi(f))$ for all $f \in A$.

(9) We show first that $C_n(h) \le C_{n'}(h)$ for $n \le n'$. Let (G'_ρ) denote the generic h'-computation on $X_1, \ldots, X_{n'}$, and (G_ρ) be the generic h-computation on $X_1, \ldots, X_n$. The substitution φ which maps $a_{\rho\sigma j}$ to 0 for all $-n'+n < j \le 0$ and which leaves all other a-variables and the b-variables invariant sends G'_ρ to G_ρ (up to a renaming of the variables). This can be proven by induction on $\rho > 0$. Note that $G'_\rho \mapsto G_\rho$ implies $G'_{\rho 0} \mapsto G_{\rho 0}$ by our general observation, as φ fixes the X-variables. By additionally requiring $c_\rho \mapsto 0$ for either $-n' + n < \rho \le 0$ or $\rho > n$ we get

$$
\varphi\big(\textstyle\sum_{\rho=-n'}^{n'} c_\rho G'_\rho\big) = \sum_{\rho=-n}^{n} c_\rho G_\rho \;.
$$

Since φ leaves the X-variables invariant, it commutes with taking homogeneous parts with respect to the X-variables. Thus $\varphi(C_{n'}(h')) = C_n(h)$.

A slight modification of the reasoning before yields a substitution φ which leaves the X-variables invariant and such that

$$\varphi\big(C_{n'}(h')(Z', \overline{X}_1 \ldots, \overline{X}_{n'-n}, X_1, \ldots, X_n)\big) = C_n(h)(Z, X_1, \ldots, X_n) \ ,$$

where Z', Z stand for the corresponding a, b, c-variables. This implies by our general observation

$$\begin{aligned}
\varphi\big(C_{n'}(h')(Z', \overline{X}_1 \ldots, \overline{X}_{n'-n}, X_1, \ldots, X_u, e_{u+1}, \ldots, e_n)\big) \\
= \quad C_n(h)(Z, X_1, \ldots, X_u, e_{u+1}, \ldots, e_n)
\end{aligned}$$

for $0 \le u \le n$ and $e_i \in \{0, 1\}$. If we extend φ by sending $d_{n'-n+u}$ to d_u for $0 \le u \le n$ and mapping the remaining d's to zero, we get $\varphi(D_{n'}(h')) = D_n(h)$. Hence $D_n(h) \le D_{n'}(h')$.

Assume now $h \le h'$ and let (G_ρ) and (G'_ρ) be the corresponding generic computations on $X_1, \ldots, X_n$. It is not hard to see that there exists a substitution which only changes the a-variables and that maps all G'_ρ to G_ρ. From this one concludes as above that $C_n(h) \le C_n(h')$ and $D_n(h) \le D_n(h')$. $\quad\square$

We will interpret families of polynomials $(f_{m,n})$ with double indices as sequences of polynomials by enumerating pairs $(m, n) \in \mathbb{N}^2$ according to $(m, n) \mapsto m + (m + n)(m + n + 1)/2$.

We can now state the first result of this section.

Theorem 5.31 *Let $h = (h_n)$ be a p-family such that any h_n contains the multiplication. Then the double-indexed families $(C_n(h_m))$ and $(D_n(h_m))$ are VP^h-complete, resp. VNP^h-complete with regard to p-projection.*

Proof. By Lemma 5.30(5) both families $(C_n(h_m))$ and $(D_n(h_m))$ are p-families. Part (f) of that lemma implies that $(C_n(h_m))$ is p-computable relative to h.

Assume $(f_j) \in \mathrm{VP}^h$, where f_j is a polynomial in $u(j)$ variables. By definition, there exists a p-bounded function $m \colon \mathbb{N} \to \mathbb{N}$ such that $\mathcal{L}^{h_{m(j)}}(f_j)$ is p-bounded in j. Let $n(j)$ denote the maximum of $u(j)$, $\deg f_j$, and $\mathcal{L}^{h_{m(j)}}(f_j)$. It is clear that $n(j)$ is p-bounded in j. We denote the a, b, c-variables in $C_{n(j)}(h_{m(j)})$ by $Z_1, \ldots Z_{w(j)}$. From Lemma 5.30(7) it follows that $f_j = C_{n(j)}(h_{m(j)})(z, X)$ for a suitable choice of $z \in k^{w(j)}$. Thus f_j is a projection of $C_{n(j)}(h_{m(j)})$, and we have proved the VP^h-completeness of $(C_n(h_m))$.

Let now (q_j) be a p-definable family relative to h, say

$$q_j(X_1, \ldots, X_{v(j)}) = \sum_{e \in \{0,1\}^{u(j)-v(j)}} f_j(X_1, \ldots, X_{v(j)}, e_{v(j)+1}, \ldots, e_{u(j)}) \ ,$$

where the family (f_j) is p-computable relative to h. From before we know that for each j

$$f_j(X) = C_{n(j)}(h_{m(j)})(z, X)$$

for p-bounded $m(j), n(j)$ and some $z \in k^{w(j)}$. Therefore, we see that q_j can be obtained from $D_{n(j)}(h_{m(j)})$ by the substitution $d_{v(j)} \mapsto 1$, $d_v \mapsto 0$ for $d \neq v(j)$, and $Z_i \mapsto z_i$ for all i. This shows that (q_j) is a p-projection of $(D_n(h_m))$.

It remains to prove that $(D_n(h_m))$ is p-definable relative to h. Let us abbreviate the a, b, c-variables occuring in $C_n(h_m)$ by $Z_1, \ldots, Z_w$ and define

$$g_{m,n} := \sum_{v=0}^{n} d_v E_1 \cdots E_v \, C_n(h_m)(Z, X_1, \ldots, X_v, E_{v+1}, \ldots, E_n) \ ,$$

where $E_1, \ldots, E_n$ are new indeterminates. As $(C_n(h_m))$ is p-computable relative to h, so is $(g_{m,n})$. The following equality

$$D_n(h_m) = \sum_{e \in \{0,1\}^n} g_{m,n}(Z, X_1, \ldots, X_n, e_1, \ldots, e_n)$$

proves that indeed $(D_n(h_m)) \in \mathrm{VNP}^h$. $\qquad\square$

We recall that a p-family (h_n) is called monotone, iff h_n is a projection of h_{n+1} for all n.

Corollary 5.32   (1) *If h is monotone and h_0 contains the multiplication, then $(C_n(h_n))$ and $(D_n(h_n))$ are VP^h-complete, resp. VNP^h-complete with respect to p-projection.*

(2) *For any p-family h there exist VP^h-complete and VNP^h-complete families with respect to p-projection.*

Proof. (1) This is an immediate consequence of Thm. 5.31 and the monotonicity of $(C_n(h))$ and $(D_n(h))$ expressed in Lemma 5.30(9).

(2) Let $h = (h_m)$ be any p-family and U, V, W be new indeterminates. The polynomials $\tilde{h}_m := U h_m + (1 - U)VW$ contain the multiplication and the p-family $\tilde{h} = (\tilde{h}_m)$ satisfies $\mathrm{VP}^{\tilde{h}} = \mathrm{VP}^h$. Now apply Thm. 5.31. $\qquad\square$

The next result is inspired by Baker, Gill, and Solovay [3]. Its proof is based on Thm. 5.31 combined with some diagonalization argument.

Theorem 5.33 *There exists a p-family h such that $\mathrm{VP}^h = \mathrm{VNP}^h$.*

Proof. First note the following: By Lemma 5.30(5) the degree as well as the number of variables of $D_n(h)$ are bounded from above by $p(n) := 2n^4 + 5n + 2$, provided the number of variables and the degree of the polynomial h are bounded by n.

By induction, we are going to construct a monotone sequence of polynomials $h = (h_n)$ such that the number of variables as well as the degree of h_n are bounded from above by n for $n \geq 2$. We define $m_i := i$ and $h_i := X_1 X_2$ for $i \leq 2$. Assume we have already constructed $h_0, \ldots, h_{m_t}$

$(t \geq 2)$. We set $m_{t+1} := p(m_t)$ and define $h_j := h_{m_t}$ for $m_t < j < m_{t+1}$ and put $h_{m_{t+1}} := D_{m_t}(h_{m_t})$. By Lemma 5.30(8) we have $h_{m_t} \leq h_{m_{t+1}}$ which guarantees the monotonicity. Moreover, the degree and the number of variables of $h_{m_{t+1}}$ are bounded by $m_{t+1} = p(m_t)$.

We claim that $(D_n(h_n))$ is a p-projection of h. In fact, let $n \geq 2$ be given, say $m_t \leq n < m_{t+1}$. The monotonicity of $D_n(h)$ expressed in Lemma 5.30(9) implies that

$$D_n(h_n) \leq D_{m_{t+1}}(h_{m_{t+1}}) = h_{m_{t+2}} .$$

But $m_{t+2} = p(p(m_t)) \leq p(p(n))$ and the composition of p with itself is clearly p-bounded. This shows the claim.

On the other hand, we know from Cor. 5.32(1) that $(D_n(h_n))$ is VNP^h-complete. As $(D_n(h_n))$ is also contained in VP^h, it follows that $\mathrm{VP}^h = \mathrm{VNP}^h$.

$\square$

Up to now we have not succeeded in establishing a p-family h such that $\mathrm{VP}^h \neq \mathrm{VNP}^h$. A promising approach for this is as follows (compare Bennett and Gill [8]). For each n choose independently $h_n \in k[X_1, \ldots, X_n]$ of degree at most n *at random* according to some probability distribution. Since the classes VP^h and VNP^h are invariant under finite variation of h, the event $\mathcal{E} = \{h \mid \mathrm{VP}^h \neq \mathrm{VNP}^h\}$ is a so-called tail event. Kolmogorov's zero-one law (cf. Feller [35, Chap. 4]) implies therefore that $\mathrm{Prob}(\mathcal{E}) \in \{0, 1\}$.

Conjecture 5.1 If the h_n are chosen with independent $0, 1$-coefficients, then we have $\mathrm{Prob}(\mathcal{E}) = 1$.

6

Fast Evaluation of Representations of General Linear Groups

We describe a fast algorithm to evaluate irreducible rational matrix representations of complex general linear groups GL_m with respect to a symmetry adapted basis (Gelfand-Tsetlin basis). We complement this by a lower bound, which shows that our algorithm is optimal up to a factor m^2 with regard to nonscalar complexity. Our algorithm can be used for the fast evaluation of special functions: for instance, we obtain an $O(\ell \log \ell)$ algorithm to evaluate all associated Legendre functions of order ℓ. The results of this chapter are from Bürgisser [17].

6.1   Description of the Problem

The theory of representations of Lie groups has countless applications in mathematics and physics. In particular, it is an indispensable tool of quantum mechanics. Let $D \colon G \to GL_d$ be an irreducible (finite dimensional) continuous representation of the Lie group G. How fast can we compute the representation matrix $D(g)$ for a given $g \in G$? This question contains the problem of the efficient evaluation of special functions and orthogonal polynomials, which can be interpreted as matrix entries of some suitable representation D (see Vilenkin and Klimyk [115]).

In this chapter, we investigate the above question for the complex general linear groups $G = GL_m$. This includes the case of the unitary groups $U(m)$, which are of particular importance for physics, since all the continuous irreducible representations of $U(m)$ can be obtained from the rational irreducible ones of GL_m by restriction.

It is well known that the irreducible polynomial representations of the general linear group GL_m can be labeled by decreasing sequences $\lambda \in \mathbb{N}^m$ (cf. Boerner [13]). With respect to some chosen basis, such a representation is a group morphism

$$D_\lambda : GL_m \longrightarrow GL_{d_\lambda}, \ A \mapsto D_\lambda(A) \ ,$$

and the entries of D_λ are homogeneous polynomials of degree $|\lambda| := \sum_i \lambda_i$ in the entries of A. The matrix $D_\lambda(A)$ is usually called an *invariant matrix* when the entries of A are interpreted as indeterminates. Littlewood [76, 77] found an explicit construction of invariant matrices. He obtained formulas for the polynomial entries of $D_\lambda(A)$ in terms of representations of the symmetric groups. Grabmeier and Kerber [46] gave a modern derivation of such formulas, and based on this, they designed an algorithm for computing invariant matrices. This algorithm in fact computes a sparse representation of the polynomial entries of the invariant matrix. However, the example of the $m \times m$ determinant already shows that this approach cannot be efficient for evaluating $D_\lambda(A)$ at specific entries $A \in \mathrm{GL}_m$ for large m: the determinant polynomial has $m!$ terms, but it can be computed with only $O(m^3)$ arithmetic operations using Gaussian elimination.

As unitary representations of $\mathrm{U}(m)$ are important in quantum mechanics, it is not astonishing that various explicit constructions of representations have been developed by physicists. (See for instance the books by Biedenharn and Louck [9], where the entries of invariant matrices are called boson polynomials.)

Gelfand and Tsetlin [44] derived explicit expressions for $D_\lambda(A)$ for generators A of GL_m with respect to bases adapted to the chain of subgroups

$$\mathrm{GL}_m > \mathrm{GL}_{m-1} \times \mathbb{C}^\times > \mathrm{GL}_{m-2} \times (\mathbb{C}^\times)^2 > \cdots > (\mathbb{C}^\times)^m \ .$$

Such symmetry adapted bases are also called *Gelfand-Tsetlin bases*. A very detailed account of this can be found in the book by Vilenkin and Klimyk [115, Chap. 18]; in Sect. 6.2 we just present the most basic facts.

The main result of this chapter (Thm. 6.6) is an efficient algorithm to evaluate the map D_λ with respect to a Gelfand-Tsetlin basis. It has a nonscalar cost roughly proportional to $m^2 d_\lambda$. Besides making optimal use of the symmetry (Schur's lemma), our algorithm uses several auxiliary algorithms: an efficient transformation of matrices with the blockstructure of a Jordan block to a direct sum of Jordan blocks, as well as the fast FFT-based multiplication of Toeplitz matrices with vectors. These auxiliary algorithms are described in Sect. 6.3. We remark that our algorithm was inspired by Clausen's fast Fourier transform [24] for the symmetric groups, as well as Maslen's extension to compact Lie groups, see the survey [81]. Their techniques also rely heavily on symmetry adapted bases.

In Sect. 6.5 we complement our algorithmic result by proving that d_λ nonscalar operations are indeed necessary for the computation of $D_\lambda(A)v$. This is easily obtained by combining Burnside's theorem with the dimension bound of algebraic complexity. It shows that our algorithm is optimal up to a factor of m^2 with respect to nonscalar complexity.

Our algorithm provides already in the special case of GL_2 results which are, to our best knowledge, new. We obtain a fast rational $O(\ell \log \ell)$ algorithm for computing all the associated Legendre functions $P_\ell^\mu(\cos \theta)$, $|\mu| \le \ell$, of degree ℓ from $\cos \theta$ and $\sin \theta$. (See Sect. 6.6.)

6.2 Preliminaries on Representations of GL_m

We collect first some facts about the representations of the general linear group GL_m (see [13, 37]). Consider GL_{m-1} as the subgroup of GL_m fixing the last canonical basis vector e_m. A polynomial representation $D_\lambda \colon \mathrm{GL}_m \to \mathrm{GL}(V)$ with highest weight $\lambda \in \mathbb{N}^m$ restricted to GL_{m-1} splits according to the branching rule [13, V§6] into a direct sum of representations with highest weights $\mu \in \mathbb{N}^{m-1}$, which satisfy the betweenness conditions $\lambda_j \geq \mu_j \geq \lambda_{j+1}$ for $1 \leq j < m$. It is important that each representation corresponding to such μ occurs with *multiplicity one*. Thus the decomposition of V into corresponding submodules V_μ is unique. We note that this is also the decomposition of V restricted to the subgroup $\mathrm{GL}_{m-1} \times \mathbb{C}^\times$, as the diagonal matrix $\mathrm{diag}(1,\ldots,1,t)$ operates on V_μ by multiplication with $t^{|\lambda|-|\mu|}$. We recall that a vector $v \in V$ is said to be of weight $w \in \mathbb{N}^m$ iff $D_\lambda(\mathrm{diag}(t_1,\ldots,t_m))v = t_1^{w_1}\cdots t_m^{w_m}v$.

If we restrict the representation D_λ successively according to the chain of subgroups

$$(6.1)\qquad \mathrm{GL}_m > \mathrm{GL}_{m-1} \times \mathbb{C}^\times > \mathrm{GL}_{m-2} \times (\mathbb{C}^\times)^2 > \cdots > (\mathbb{C}^\times)^m \ ,$$

we finally end up with a decomposition of V into one-dimensional subspaces of weight vectors. This decomposition is unique, and bases of V adapted to this decomposition are called *Gelfand-Tsetlin bases*. Thus Gelfand-Tsetlin bases are uniquely determined up to a permutation of the basis elements and scaling. We remark that if V is a (finite dimensional) Hilbert space and D_λ restricted to $\mathrm{U}(m)$ is unitary, then a Gelfand-Tsetlin basis can be chosen to orthonormal.

The splitting behaviour can be conveniently visualized by a layered graph $G(\lambda)$, whose nodes on level n ($1 \leq n \leq m$) are the occuring irreducible representations of V restricted to $\mathrm{GL}_n \times (\mathbb{C}^\times)^{m-n}$. These nodes can thus be uniquely described by pairs (ν, w), where $\nu \in \mathbb{N}^n$ is a partition and $w \in \mathbb{N}^{m-n}$ satisfies $|\nu| + |w| = |\lambda|$. A node on level n is connected in the graph $G(\lambda)$ with a node on level $n-1$ if the latter appears in the decomposition of the former upon restriction to $\mathrm{GL}_{n-1} \times (\mathbb{C}^\times)^{m-n+1}$ (see Fig. 6.1).

The number of paths in $G(\lambda)$ between a node $x = (\nu, w)$ at level n and a node $x' = (\nu', w')$ at level $n' < n$ is just the multiplicity with which x' occurs in x when restricted to $\mathrm{GL}_{n'} \times (\mathbb{C}^\times)^{m-n'}$. We denote this multiplicity by $\mathrm{mult}(x, x')$. We call the maximum of $\mathrm{mult}(x, x')$ taken over all pair of nodes *two levels apart* the *multiplicity* $\mathrm{mult}(\lambda)$ of the highest weight λ. (In Fig. 6.1 we have $\mathrm{mult}(\lambda) = 2$.)

The vectors of a Gelfand-Tsetlin basis can be labeled by paths in $G(\lambda)$ going from the top node λ to a node at level one. Such paths can be encoded as semistandard tableaus: for instance, in Fig. 6.1 we have two vectors of weight $(1,1,1)$ corresponding to the tableaus $\begin{smallmatrix}1&2\\3\end{smallmatrix}$ and $\begin{smallmatrix}1&3\\2\end{smallmatrix}$. The quantity $\mathrm{mult}(x, x')$ can thus be alternatively described as the number of semistandard tableaus of the skew diagram $\nu \setminus \nu'$ in which j occurs exactly w'_j times ($n' < j \leq n$).

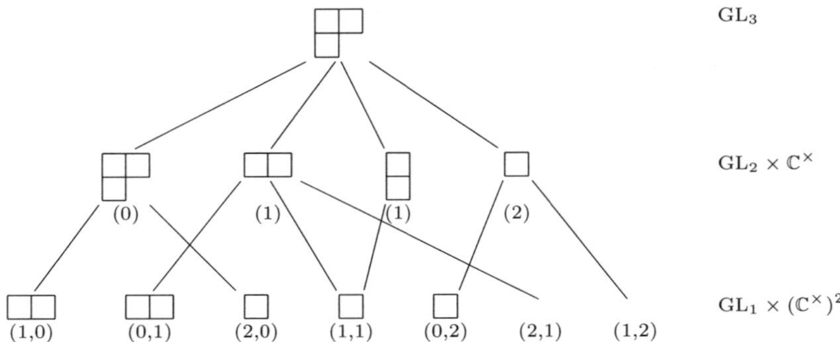

Figure 6.1: The graph $G(\lambda)$ for $\lambda = (2,1,0)$, $m = 3$.

These considerations also imply that the dimension d_λ equals the number of semistandard tableaus of the diagram λ. Moreover, the number of vectors of weight $(1, \ldots, 1)$ in a Gelfand-Tsetlin basis corresponding to λ equals the number s_λ of standard tableaus on the diagram of λ.

Remark 6.1   (1) We have $d_\lambda \geq m$ if $\lambda = (\lambda_1, \ldots, \lambda_m) \neq (m, \ldots, m)$.

(2) We have $\mathrm{mult}(\lambda) \leq 2$ for *hook partitions* $\lambda = (m - i, 1, \ldots, 1)$.

By suitable scaling and ordering of the vectors of a Gelfand-Tsetlin basis, we can obtain a basis of V, which is adapted to the chain (6.1) of subgroups in the following strong sense: the corresponding matrix representation D of GL_m satisfies for all n:

(1) The restriction $D \downarrow \mathrm{GL}_n \times (\mathbb{C}^\times)^{m-n}$ is *equal* to a direct sum of matrix representations of this subgroup.

(2) Equivalent irreducible constituents of $D \downarrow \mathrm{GL}_n \times (\mathbb{C}^\times)^{m-n}$ are *equal*.

These properties are crucial for our computational purpose. For convenience, we will call such adapted bases also *Gelfand-Tsetlin bases*.

Consider now the matrix $B_{i,j}(t) \in \mathrm{GL}_m$ with entries 1 in the diagonal, entry t at position (i, j), and having entries 0 elsewhere $(i \neq j)$. Let D: $\mathrm{GL}_m \to \mathrm{GL}(V)$ be a rational representation. Then $F = F_{i,j}: \mathbb{C} \to \mathrm{GL}(V)$, $t \mapsto D(B_{i,j}(t))$ is a one-parameter subgroup: we have $F(s + t) = F(s)F(t)$ for $s, t \in \mathbb{C}$. Hence $F'(t) = F'(0)F(t)$ and therefore $F(t) = e^{tF'(0)}$. (Note that $F'(0)$ must be nilpotent.) Let $\epsilon_i \in \mathbb{N}^m$ denote the basis vector having components 0 except at position i, where the component equals 1.

Lemma 6.2 $F'_{i,j}(0)$ *maps a vector of weight* $w \in \mathbb{Z}^m$ *into one of weight* $w + \epsilon_i - \epsilon_j$.

Proof. For a fixed vector v of weight w we may write

$$F(t)(v) = \sum_{s \geq 0} t^s u_s$$

with vectors $u_s \in V$. We are going to show that u_s must be a vector of weight $w + s(\epsilon_i - \epsilon_j)$. Then we are done, since $F'(0)(v) = u_1$.

If $g = \mathrm{diag}(g_i)$ is a diagonal matrix, we have $gB_{i,j}(t)g^{-1} = B_{i,j}(g_i t g_j^{-1})$. This implies

$$\sum_{s \geq 0} t^s D(g)(u_s) = D(g)F(t)(v) \quad = \quad F(g_i t g_j^{-1})D(g)(v)$$

$$= \quad g_1^{w_1} \cdots g_m^{w_m} \cdot \sum_{s \geq 0} (g_i t g_j^{-1})^s u_s \ ,$$

as $D(g)(v) = g_1^{w_1} \cdots g_m^{w_m} v$. By comparing the coefficients of t, we see that u_s is indeed a vector of weight $w + s(\epsilon_i - \epsilon_j)$. $\qquad \square$

6.3   Auxiliary Fast Linear Algebra Algorithms

We present some auxiliary algorithms, which we will need as subroutines in our algorithm for evaluating representations.

The first one is a variant of Gaussian elimination.

Lemma 6.3 *Any matrix $A \in \mathrm{GL}_m$ can be factored as $A = A_N A_{N-1} \cdots A_1 \Delta$, where $N \leq 2m^2$, Δ is a diagonal matrix, and all A_i are elementary matrices of the form $B_{n-1,n}(t)$ or $B_{n,n-1}(t)$. Moreover, such a decomposition can be computed with $O(m^3)$ arithmetic operations.*

Proof. Recall that multiplying a matrix from the left by $B_{i,j}(t)$ has the effect of adding the t-fold of the jth row to the ith row. Also note the following: suppose the jth row of A equals zero. Then multiplying A from the left by $B_{i,j}(-1)B_{j,i}(1)$ has the effect of interchanging the jth row with the ith row.

By a sequence of elementary row operations affecting only *neighbouring* rows, we can transfom a given invertible matrix $A \in \mathrm{GL}_m$ into diagonal form Δ. Hereby, we first take care of the first column of A by working up from the bottom row, then we deal with the second column in a similar way and so forth. Let us illustrate the procedure in the case $m = 4$. The symbol $C_{i,j}$ denotes either $B_{i,j}(-1)B_{j,i}(1)$, or $B_{i,j}(t)$ for some $t \in \mathbb{C}$. We can obtain a decomposition of the form

$$(C_{3,4}C_{2,3}C_{1,2})(C_{2,3}C_{1,2}C_{4,3})(C_{1,2}C_{3,2}C_{4,3})(C_{2,1}C_{3,2}C_{4,3})\, A = \Delta \ ,$$

where Δ is a diagonal matrix. (The parenthesis indicate the treatment of the columns.) The number of occuring $C_{i,j}$ matrices equals $m(m-1)$ in the general situation. Moreover, the $C_{i,j}$ and Δ can be computed with $O(m^3)$ arithmetic operations from A. This proves the lemma. $\qquad \square$

The next result is well-known and relies on the fast Fourier transform (see for instance [21, Cor. 13.13]).

Proposition 6.4 *Suppose $J \in \mathbb{C}^{r \times r}$ is a nilpotent Jordan block. Then e^{tJ} is a Toeplitz matrix. Thus $e^{tJ}u$ can be computed from $t \in \mathbb{C}$ and $u \in \mathbb{C}^r$ with $O(r \log r)$ arithmetic operations. For this, $O(r)$ nonscalar operations are sufficient.*

We remark that the computation of $e^{tJ}u$ is equivalent to the task of evaluating the polynomial $f(T) = \sum_{j=0}^{r-1} \frac{u_j}{j!} T^j$ and all its derivatives at $t \in \mathbb{C}$.

Part (a) of the following lemma shows that a matrix having the block-structure of a Jordan block can be efficiently transformed to a direct sum of Jordan blocks. Part (b) follows then easily with Prop. 6.4.

Lemma 6.5 *Let $M \in \mathbb{C}^{m \times m}$ be a matrix decomposed into r^2 blocks $M_{ij} \in \mathbb{C}^{m_i \times m_j}$, $m = m_1 + \ldots + m_r$. Suppose that all block entries outside the lower diagonal are zero, that is, $M_{ij} = 0$ if $i \neq j + 1$. Then the following is true:*

(1) *There is an invertible block diagonal matrix $S = [S_{ij}] \in \mathbb{C}^{m \times m}$ ($S_{ij} \in \mathbb{C}^{m_i \times m_j}$, $S_{ij} = 0$ for $i \neq j$) and a permutation matrix P such that the $PSMS^{-1}P^{-1}$ is a direct sum of nilpotent Jordan blocks of size at most r.*

(2) *The product $e^{tM}u$ can be computed with $O(\sum_{\rho=1}^{r} m_\rho^2 + m \log r)$ arithmetic operations from $t \in \mathbb{C}$ and $u \in \mathbb{C}^m$.*

Proof.
(1) It is convenient to take a coordinate-free point of view. Let $V = V_1 \oplus \ldots \oplus V_r$ be a decomposition of vector spaces together with linear maps $\varphi_\rho \colon V_\rho \to V_{\rho+1}$ for $1 \leq \rho < r$. Let $\varphi \colon V \to V$ be the linear map satisfying $\varphi(v) = \varphi_\rho(v)$ if $v \in V_\rho$, $\rho < r$, and $\varphi(v) = 0$ if $v \in V_r$. (Note that φ and φ_ρ are coordinate-free versions of M and $M_{\rho+1,\rho}$, respectively.) By a φ-chain of length t we understand an ordered set of vectors $\{v_1, \ldots, v_t\}$ such that $\varphi(v_\tau) = v_{\tau+1}$ for $1 \leq \tau < t$, and $\varphi(v_t) = 0$. (φ-chains correspond to nilpotent Jordan blocks.) The claim of (a) amounts to showing the existence of a basis E_ρ for each V_ρ such that the basis $E_1 \cup \ldots \cup E_r$ of V is a disjoint union of φ-chains of length at most r.

For $1 \leq \rho \leq \sigma < r$ let $V_{\rho,\sigma}$ denote the kernel of the composition

$$\varphi_\sigma \circ \cdots \circ \varphi_{\rho+1} \circ \varphi_\rho \colon V_\rho \to V_{\sigma+1}$$

and set $V_{\rho,r} = V_\rho$ for $1 \leq \rho \leq r$. Note that $V_{\rho,\sigma} \subseteq V_{\rho,\sigma+1}$ and $\varphi_\rho^{-1}(V_{\rho+1,\sigma}) = V_{\rho,\sigma}$. In particular, $\varphi_\rho(V_{\rho,\sigma}) \subseteq V_{\rho+1,\sigma}$.

Choose subsets $E_{1,\sigma} \subseteq V_{1,\sigma}$ such that $E_{1,1} \cup \ldots, E_{1,\sigma}$ is a basis of $V_{1,\sigma}$ for all $1 \leq \sigma \leq r$. (This means that $E_{1,1} \cup \ldots \cup E_{1,r}$ is a basis of V_1 adapted to the flag $V_{1,1} \subseteq \ldots \subseteq V_{1,r}$ of subspaces.)

By induction on $\rho = 2, \ldots, r$ we are going to construct finite subsets $E_{\rho,\sigma} \subseteq V_{\rho,\sigma}$ satisfying the following conditions for all $\rho \leq \sigma \leq r$:

$(i)_\rho$ $E_{\rho,\rho} \cup \ldots \cup E_{\rho,\sigma}$ is a basis of $V_{\rho,\sigma}$,

$(ii)_\rho$ $\varphi_{\rho-1}(E_{\rho-1,\sigma}) \subseteq E_{\rho,\sigma}$.

Assume we have already constructed the $E_{\rho-1,\sigma}$ satisfying the conditions $(i)_{\rho-1}$ and $(ii)_{\rho-1}$. We claim that the subset $\varphi_{\rho-1}(E_{\rho-1,\sigma}) \subseteq V_{\rho,\sigma}$ is linearly independent modulo $V_{\rho,\sigma-1}$, provided $\rho \leq \sigma$. Indeed, if we had a nontrivial linear combination

$$\sum_{v \in E_{\rho-1,\sigma}} \lambda_v \varphi_{\rho-1}(v) \in V_{\rho,\sigma-1} \ ,$$

then we would obtain

$$\sum_{v \in E_{\rho-1,\sigma}} \lambda_v v \in \varphi_{\rho-1}^{-1}(V_{\rho,\sigma-1}) = V_{\rho-1,\sigma-1} \ .$$

By our inductive assumption $(i)_{\rho-1}$, the set $E_{\rho-1,\rho-1} \cup \ldots \cup E_{\rho-1,\sigma-1}$ is a basis of $V_{\rho-1,\sigma-1}$ and $E_{\rho-1,\rho-1} \cup \ldots \cup E_{\rho-1,\sigma}$ is linearly independent. This is a contradiction! We may now choose subsets $E_{\rho,\sigma}$ containing $\varphi_{\rho-1}(E_{\rho-1,\sigma})$ such that $E_{\rho,\sigma}$ is linearly independent modulo $V_{\rho,\sigma-1}$. Then the conditions $(i)_\rho$ and $(ii)_\rho$ are satisfied.

We have now constructed a basis $E_\rho := E_{\rho,\rho} \cup \ldots \cup E_{\rho,r}$ for each of the spaces V_ρ. We write the basis $E := E_1 \cup \ldots \cup E_r$ as the disjoint union of the subsets

$$F := \bigcup_{\rho < \sigma} E_{\rho,\sigma} \quad \text{and} \quad G := \bigcup_\rho E_{\rho,\rho} \ .$$

By our construction, φ induces an injective map from F to $F \cup G$, and we have $\varphi(G) = \{0\}$. This abstract property easily implies that E is a disjoint union of φ-chains. It is obvious that the length of these chains cannot be bigger than r.

(2) Let $PSMS^{-1}P^{-1} = J = J_1 \oplus \ldots \oplus J_s$ be as in the statement of part (1) of this lemma. The J_σ are nilpotent Jordan blocks of size $r_\sigma \leq r$. We have

$$e^{tM} = S^{-1}P^{-1}e^{tJ}PS, \ e^{tJ} = \bigoplus_{\sigma=1}^{s} e^{tJ_\sigma} \ .$$

We can compute $u' = Su$ from a given $u \in \mathbb{C}^m$ with $O(\sum_{\rho=1}^{r} m_\rho^2)$ arithmetic operations, since S is a block diagonal matrix. The vectors $u'_\sigma \in \mathbb{C}^{r_\sigma}$ satisfying $\oplus_{\sigma=1}^{s} u'_\sigma = Pu'$ are obtained without further arithmetic operations. By Prop. 6.4, we can compute each of the products $e^{tJ_\sigma} u'_\sigma$ with $O(r_\sigma \log r_\sigma)$ arithmetic operations. Thus we get $u'' := e^{tJ} \oplus_\sigma u'_\sigma$ from the u'_σ using $O(\sum_{\sigma=1}^{s} r_\sigma \log r_\sigma) \leq O(m \log r)$ operations. Summarizing, we have computed $e^{tM} u$ from t and u using $O(\sum_{\rho=1}^{r} m_\rho^2 + m \log r)$ arithmetic operations. $\qquad\square$

6.4 An Algorithm for Evaluating Representations

The main result of this chapter is expressed in the following theorem. We call $\lambda \in \mathbb{N}^m$ constant iff its components are all equal.

Theorem 6.6 *Let D_λ be the matrix representation of* GL_m *with respect to a Gelfand-Tsetlin basis with highest weight $\lambda \in \mathbb{N}^m$ We suppose that λ is not constant. Then the map*

$$\mathrm{GL}_m \times \mathbb{C}^{d_\lambda} \longrightarrow \mathbb{C}^{d_\lambda}, \ (A, v) \mapsto D_\lambda(A)v$$

can be computed with $O(m^2(\mathrm{mult}(\lambda) + \log |\lambda|)\, d_\lambda)$ arithmetic operations. The nonscalar complexity is bounded by $O(m^2 d_\lambda + m\lambda_1)$.

Remark 6.7 (1) We assume exact arithmetic of complex numbers.

(2) In the above upper bound the cost for *constructing* the algorithm (preconditioning) is not taken into account. However, the explicit formulas in Vilenkin and Klimyk [115, Chap. 18] for invariant matrices evaluated at special generators of GL_m suggest that this can also be done very efficiently.

(3) We need the assumption that λ is not constant. Otherwise, the determinant could be evaluated with $O(m^2 \log m)$ operations (take $\lambda = (1, \ldots, 1)$).

Corollary 6.8 *The invariant matrix with respect to a nonconstant $\lambda \in \mathbb{N}^m$ and a Gelfand-Tsetlin basis can be evaluated at a matrix $A \in \mathrm{GL}_m$ with a number of $O(m^2(\mathrm{mult}(\lambda) + \log |\lambda|)\, d_\lambda^2)$ arithmetic operations.*

Proof. (of Thm. 6.6) We first factor the given matrix $A \in \mathrm{GL}_m$ according to Lemma 6.3 as $A = A_N A_{N-1} \cdots A_1 \Delta$. Note that the cost for doing this is dominated by $O(m^2 d_\lambda)$, since $d_\lambda \geq m$ for a nonconstant $\lambda \in \mathbb{N}^m$ by Remark 6.1. We therefore have $D_\lambda(A) = D_\lambda(A_N)D_\lambda(A_{N-1}) \cdots D_\lambda(A_1)D_\lambda(\Delta)$. For given $v \in \mathbb{C}^{d_\lambda}$ we first compute $v_0 = D_\lambda(\Delta)v$ and then sucessively $v_i = D_\lambda(A_i)v_{i-1}$ for $1 \leq i \leq N$. Obviously, $v_t = D_\lambda(A)v$.

As a Gelfand-Tsetlin basis consists of weight vectors, the matrix $D_\lambda(\Delta)$ is diagonal with entries $t_1^{w_1} \cdots t_m^{w_m}$, where $w_i \leq \lambda_1$. Thus $D_\lambda(\Delta)v$ can be certainly computed with $O(m\, d_\lambda \log \lambda_1) \leq O(m\, d_\lambda \log |\lambda|)$ arithmetic operations. (Note that $2 \log w_i$ multiplications are sufficient to obtain $t_i^{w_i}$.)

It remains to show that we can compute each of the products $D_\lambda(A_i)v_i$ with $O((\mathrm{mult}(\lambda) + \log |\lambda|)d_\lambda)$ arithmetic operations. We assume that $A_i = B_{n-1,n}(t)$, the case of $A_i = B_{n,n-1}(t)$ being analogous. Let us write $V = \mathbb{C}^{d_\lambda}$ and interpret V as a GL_m-module via D_λ. Recall the graph $G(\lambda)$ introduced in Sect. 6.2, which describes the splitting behaviour of V. We have

$$(6.2) \qquad\qquad V \downarrow \mathrm{GL}_n \times (\mathbb{C}^\times)^{m-n} = \bigoplus_x \bigoplus_{j=1}^{f(x)} V_{x,j} \ ,$$

where the first sum is over all nodes x of $G(\lambda)$ at level n, $f(x)$ equals $\mathrm{mult}(\lambda, x)$, and $V_{x,j}$ is an irreducible $\mathrm{GL}_n \times (\mathbb{C}^\times)^{m-n}$-module of type x. Because of the symmetry adaptation, the decomposition (6.2) is compatible with our Gelfand-Tsetlin basis (i.e., subsets of this basis form a basis of each $V_{x,j}$). As A_i is contained in GL_n, $D_\lambda(A_i)$ decomposes according to (6.2) into

$$D_\lambda(A_i) = \oplus_x \oplus_j D_{x,j}(A_i) \ .$$

If we write $v_i = \oplus_x \oplus_j v_{x,j}$ according to (6.2) (this decomposition can be done for free), we have $D_\lambda(A_i)v_i = \oplus_x \oplus_j D_{x,j}(A_i)v_{x,j}$. Now it is sufficient to show that each of the products $D_{x,j}(A_i)v_{x,j}$ can be computed with $O((\mathrm{mult}(\nu_x)+\log|\nu_x|)\, d_{\nu_x})$ arithmetic operations, where $x = (\nu_x, w)$, $\nu_x \in \mathbb{N}^n$, $d_{\nu_x} = \dim V_{x,j}$. Indeed, then $D_\lambda(A_i)v_i$ can be computed with a number of arithmetic operations bounded by

$$\sum_x (\mathrm{mult}(\nu_x) + \log|\nu_x|)\, f(x)d_{\nu_x} \quad \leq \quad (\mathrm{mult}(\lambda) + \log|\lambda|) \sum_x f(x)d_{\nu_x}$$
$$= \quad (\mathrm{mult}(\lambda) + \log|\lambda|)\, d_\lambda$$

up to a constant factor. By symmetry adaptation, a subset of the original Gelfand-Tsetlin basis forms a Gelfand-Tsetlin basis of $V_{x,j}$ and $D_{x,j}$ is the corresponding matrix representation. We may therefore continue our argumentation assuming $n = m$.

We have to prove now that we can compute the product $D_\lambda(B_{m-1,m}(t))v$ with $O((\mathrm{mult}(\lambda) + \log|\lambda|)\, d_\lambda)$ arithmetic operations. In order to show this, we put $F(t) := D_\lambda(B_{m-1,m}(t))$ and $\Gamma := F'(0)$. Similarly as in (6.2) we have the following decomposition

$$(6.3) \qquad V \downarrow \mathrm{GL}_{m-2} \times (\mathbb{C}^\times)^2 = \bigoplus_\nu \bigoplus_{a=0}^{|\lambda|-|\nu|} \bigoplus_{j=1}^{f(\nu,a)} V_{\nu,a,j} \ ,$$

where the first two sums are over all ν, a such that the pair $x := (\nu, w)$ with $w := (a, |\lambda| - |\nu| - a))$ is a node of $G(\lambda)$ at level $m - 2$. The irreducible $\mathrm{GL}_{m-2} \times (\mathbb{C}^\times)^2$-module $V_{\nu,a,j}$ is of type x, and $f(\nu, a) = \mathrm{mult}(\lambda, x)$. Note that $f(\nu, a) \leq \mathrm{mult}(\lambda)$.

$F(t)$ commutes with GL_{m-2}, hence so does Γ. Therefore, Γ maps the isotypical components

$$W_\nu := \bigoplus_a \bigoplus_j V_{\nu,a,j}$$

of $V\downarrow\mathrm{GL}_{m-2}$ into itself. Note that $\dim W_\nu = g(\nu)d_\nu$, where $g(\nu) := \sum_a f(\nu, a)$. From Lemma 6.2 we know that Γ maps $\oplus_j V_{\nu,a,j}$ into $\oplus_j V_{\nu,a+1,j}$. We decompose now Γ according to (6.3) into $d_\nu \times d_\nu$-matrices $\Gamma_{(\nu',a',j'),(\nu,a,j)}$. From the observations made just before we see that these matrices vanish unless $\nu' = \nu$ and $a' = a + 1$. Such a matrix affords a GL_{m-2}-module morphism $V_{\nu,a,j} \to V_{\nu,a',j'}$. On the other hand, the identity matrix affords as well a

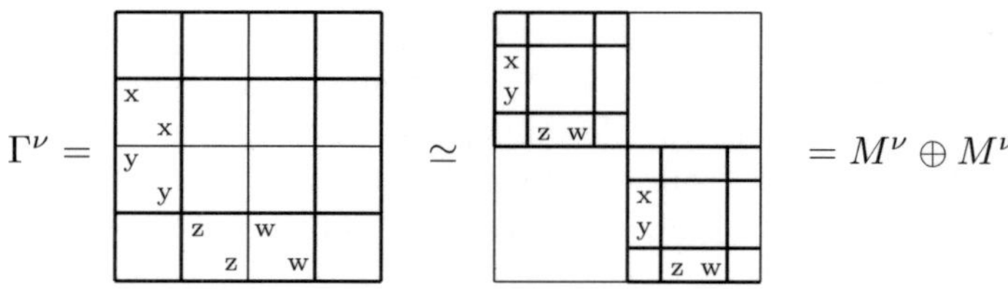

Figure 6.2: The matrix Γ^ν for $\lambda = (2,1,0,0)$, $\nu = (1,0)$. We have $f(\nu,0) = 1$, $f(\nu,1) = 2$, $f(\nu,2) = 1$, $g(\nu) = 4$, $d_\nu = 2$.

GL_{m-2}-module morphism between these spaces, since our basis is adapted to this subgroup. Schur's lemma implies therefore that $\Gamma_{(\nu,a',j'),(\nu,a,j)}$ must be a multiple of the $d_\nu \times d_\nu$ identity matrix.

Let Γ^ν denote the matrix $[\Gamma_{(\nu,a',j'),(\nu,a,j)}]_{(a',j'),(a,j)}$, that is, the matrix of Γ restricted to W_ν. It is not hard to see that Γ^ν equals, after some suitable permutation of our basis of W_ν, a direct sum of d_ν *identical copies* of a matrix $M^\nu \in \mathbb{C}^{g(\nu) \times g(\nu)}$ (see Fig. 6.2). This matrix M^ν has a decomposition into $(|\lambda| - |\nu| + 1)^2$ blocks $M^\nu_{a',a} \in \mathbb{C}^{f(\nu,a') \times f(\nu,a)}$ and all blocks outside the lower diagonal vanish: $M^\nu_{a',a} = 0$ unless $a' = a + 1$.

From Lemma 6.5(2) we know that a product $e^{tM^\nu} u$ can be computed from $t \in \mathbb{C}$ and $u \in \mathbb{C}^{g(\nu)}$ with a number of arithmetic operations bounded by

$$\sum_a f(\nu,a)^2 + g(\nu)\log(|\lambda| - |\nu| + 1) \le g(\nu)\mathrm{mult}(\lambda) + g(\nu)\log(|\lambda| + 1)$$

up to a constant factor. Therefore, a product $F(t)v$ can be computed with

$$O\left(\sum_\nu d_\nu g(\nu)\big(\mathrm{mult}(\lambda) + \log(|\lambda| + 1)\big)\right)$$

arithmetic operations, which proves the claim, as $d_\lambda = \sum_\nu d_\nu g(\nu)$.

The estimation of the nonscalar complexity is similar. $\qquad\square$

6.5 A Lower Bound

The subsequent lower bound result shows that our algorithm is optimal up to a factor of m^2 with respect to the number of nonscalar operations.

Theorem 6.9 *Let $\lambda \in \mathbb{N}^m$ be such that $|\lambda| > 1$ and let $v \in \mathbb{C}^{d_\lambda}$ be nonzero. Let D_λ denote a matrix representation of GL_m with highest weight λ. Then any arithmetic algorithm (formally, algebraic computation tree) computing the map $\mathrm{GL}_m \to \mathbb{C}^{d_\lambda}$, $A \mapsto D_\lambda(A)v$ requires at least d_λ nonscalar operations. The evaluation of the invariant matrix $\mathrm{GL}_m \to \mathrm{GL}_{d_\lambda}$, $A \mapsto D_\lambda(A)$ requires at least d_λ^2 nonscalar operations.*

Proof. The theorem of Burnside (cf. [73]) states that the linear hull of the image of D_λ equals $\mathbb{C}^{d_\lambda \times d_\lambda}$, as D_λ is irreducible. Therefore, the entries of the invariant matrix corresponding to λ are linearly independent polynomials of degree $|\lambda| > 1$. The dimension bound in [21, (4.12))] easily implies the claims.

$\square$

6.6 Fast Evaluation of Legendre Functions

Our algorithm of Sect. 6.4 can be applied to evaluate many special functions and orthogonal polynomials (Legendre, Jacobi, Gegenbauer polynomials, generalized Beta functions, etc.), since all these are matrix entries of a suitable representation of GL_m (compare [115]). We illustrate this here by the simple example of the associated Legendre functions.

For the following facts about the representations of GL_2 or $\mathrm{SU}(2)$ see [115, §6.2]. The natural operation of GL_2 on the space V of homogeneous bivariate polynomials of degree 2ℓ in the indeterminates X, Y affords an irreducible representation of highest weight $\lambda = (2\ell, 0)$ and dimension $d_\lambda = 2\ell + 1$. All irreducible representations of GL_2 are obtained in this way for $\ell \in \frac{1}{2}\mathbb{N}$. Note that $\mathrm{mult}(\lambda) = 1$. In the sequel, we will assume that $\ell \in \mathbb{N}$. The basis $(\psi_k)_{-\ell \leq k \leq \ell}$ given by the elements

$$\psi_k := \frac{X^{\ell-k} Y^{\ell+k}}{\sqrt{(\ell - k)!(\ell + k)!}}$$

is a Gelfand-Tsetlin basis having the additional property that the corresponding matrix representation $D^\ell = [D^\ell_{\mu\nu}]_{-\ell \leq \mu,\nu \leq \ell}$ restricted to $\mathrm{SU}(2)$ is unitary. Consider the special unitary matrix

$$A(\theta) := \begin{pmatrix} \cos\theta/2 & i\sin\theta/2 \\ i\sin\theta/2 & \cos\theta/2 \end{pmatrix} .$$

Thm. 6.6 yields an algorithm to compute a column of $D^\ell(A(\theta))$ from $\cos\theta/2$ and $\sin\theta/2$ with $O(\ell \log \ell)$ arithmetic operations. A closer look at this algorithm reveals that the computation may start with $\cos\theta$ and $\sin\theta$ (no square roots are necessary). Indeed, we have the matrix factorization

$$A(\theta) = \begin{pmatrix} 1 & 0 \\ a & 1 \end{pmatrix} \begin{pmatrix} 1 & b \\ 0 & 1 \end{pmatrix} \begin{pmatrix} c & 0 \\ 0 & c^{-1} \end{pmatrix} ,$$

where $a = i(1 - \cos\theta)/\sin\theta$, $2b = i\sin\theta$, and $c = \cos\theta/2$. Moreover, note that $c^{w_1}(c^{-1})^{w_2} = (c^2)^{w_1-\ell} = (\frac{1+\cos\theta}{2})^{w_1-\ell}$ for all weights (w_1, w_2) of V.

In turns out that the middle column of $D^\ell(A(\theta))$ just contains, up to some scaling, the associated Legendre functions P_ℓ^μ (cf. [115, §6.3.7 (3)]: we have for $|\mu| \leq \ell$

$$D^\ell_{-\mu,0}(A(\theta)) = i^{-\mu}\sqrt{\frac{(\ell - \mu)!}{(\ell + \mu)!}}\, P_\ell^\mu(\cos\theta) .$$

Corollary 6.10 *All the associated Legendre functions $P_\ell^\mu(\cos\theta)$, $|\mu| \le \ell$, can be computed from $\cos\theta$ and $\sin\theta$ using only $O(\ell \log \ell)$ arithmetic operations. The nonscalar complexity is bounded by $O(\ell)$.*

Remark 6.11 One can give a direct proof of Cor. 6.10 as follows. The associated Legendre functions P_ℓ^μ satisfy ($0 \le \mu \le \ell$)

$$P_\ell^\mu(x) = (-1)^\mu (1 - x^2)^{\mu/2} \frac{d^\mu P_\ell(x)}{dx^\mu} \; ,$$

where $P_\ell(x)$ denotes the Legendre polynomial of degree ℓ. (This again shows that $P_\ell^\mu(\cos\theta)$ is a polynomial in $\cos\theta$ and $\sin\theta$.) By Prop. 6.4, we can evaluate P_ℓ and all its derivatives at x with $O(\ell \log \ell)$ arithmetic operations (compare the comment following this proposition). Moreover, we have for $0 \le \mu \le \ell$ that

$$P_\ell^{-\mu}(x) = (-1)^\mu \frac{\Gamma(\ell - \mu + 1)}{\Gamma(\ell + \mu + 1)} P_\ell^\mu(x)$$

(cf. [115, §3.5.8 (10)], Γ stands for the Gamma function). This already finishes the alternative proof of Cor. 6.10.

7

The Complexity of Immanants

For studying the complexity to evaluate single entries of invariant matrices of GL_m, we investigate the complexity of immanants. These matrix functions contain the permanent and determinant as special cases. From our algorithm in Chap. 6, we deduce an algorithm to evaluate immanants, which is faster than previous ones due to Hartmann and Barvinok. Finally, we show that the problem to evaluate certain immanants corresponding to hook diagrams or rectangular diagrams is complete in Valiant's sense. The results in this chapter are taken from Bürgisser [17, 18].

7.1 Motivation and Outline of Chapter

For computing individual entries of the invariant matrix, the cost of our algorithm in Chap. 6 may appear prohibitively high, mainly because the dimension d_λ can be very large. For instance, for $\lambda = (m, 0, \ldots, 0) \in \mathbb{N}^m$, we have $d_\lambda = \binom{2m-1}{m}$, which is exponential in m. Is this inherent to the problem, or are there faster algorithms running with a number of steps polynomially bounded in m?

It is one of the goals of this chapter to give partial answers to this question. For approaching it, we do not focus on individual entries of the invariant matrix, but we study related functions having some invariant meaning. Let λ be a partition of m (or a Young diagram with m boxes). We consider the function sending $A \in GL_m$ to the sum of the diagonal entries of the invariant matrix $D_\lambda(A)$ corresponding to the weight $(1, \ldots, 1)$. This turns out to be the so-called *immanant*

$$\mathrm{im}_\lambda(A) = \sum_{\pi \in S_m} \chi_\lambda(\pi) \prod_{i=1}^m A_{i,\pi(i)} \ ,$$

of the matrix A corresponding to λ, which was introduced by Littlewood [77] (see Lemma 7.2). Here, χ_λ denotes the irreducible character of the symmetric group S_m belonging to λ (cf. Boerner [13], or James and Kerber [51]). This notion contains the permanent and determinant as special cases ($\chi_\lambda = 1$ or $\chi_\lambda = \mathrm{sgn}$).

From the viewpoint of computational complexity, determinant and permanent have, in spite of the similarity in their definition, very little in common. While there are efficient polynomial algorithms for the evaluation of the determinant, Valiant's completeness result (Thm. 2.10) indicates that the permanent cannot be computed with a polynomial number of arithmetic operations. Given this situation, it is quite natural to raise the question of the computational complexity of immanants. First upper bounds and lower bounds on this problem were obtained by Hartmann [47]. A different algorithm for computing immanants, which in some cases improves Hartmann's bound, was found by Barvinok [5].

From our algorithm for evaluating representations of GL_m, we derive in Sect. 7.2 an upper bound on the computational complexity of immanants, which improves the previous bounds due to Hartmann and Barvinok. In passing, we also show that the characters of GL_m can be evaluated extremely fast: namely (asymptotically) as fast as matrix multiplication.

The remaining part of the chapter is devoted to completeness results for certain families of immanants, which are stated in Sect. 7.3. We generalize a result of Hartmann [47] by proving that families of immanants corresponding to certain hook diagrams are VNP-complete (Thm. 7.5). Moreover, we succeed in proving that immanants corresponding to square diagrams, and more generally, to certain rectangular diagrams, are VNP-complete (Thm. 7.6), thereby solving an open problem posed by Strassen [106, Problem 14.2]. The meaning of these results is that these immanants cannot be computed with a polynomial number of arithmetic operations if Valiant's hypothesis is true.

The rest of the chapter is organized as follows. In Sect. 7.4 we derive a character formula for the symmetric group, which expresses values of characters corresponding to rectangular Young diagrams by characters of hook diagrams (Lemma 7.8). Sect. 7.5 is devoted to the proof that families of immanants are indeed p-definable. Finally, in Sect. 7.6, we provide the proofs for our completeness results. Besides Lemma 7.8, the central tool is a consequence of the Murnaghan-Nakayama rule for the characters of the symmetric groups, which allows to identify certain alternating sums of immanants corresponding to smaller partitions as a projection of a given immanant (Lemma 7.11). Using this strategy repeatedly, it is possible to obtain a permanent or Hamilton cycle polynomial (of large size) as a projection of the immanant under investigation.

7.2 Fast Evaluation of Immanants

We recall the definition of immanants. Let χ_λ denote the irreducible character of the symmetric group S_m belonging to a partition λ of m (cf. [13, 51]).

Definition 7.1 The *immanant* of a matrix $A \in \mathbb{C}^{m \times m}$ corresponding to a partition λ of m is defined as

$$\mathrm{im}_\lambda(A) = \sum_{\pi \in S_m} \chi_\lambda(\pi) \prod_{i=1}^m A_{i,\pi(i)} \ .$$

The reader may find some algebraic properties of immanants in the work of Merris [82, 83].

The following lemma reveals a connection of immanants to the representations of GL_m. It states that immanants appear naturally as the sum of the diagonal entries of invariant matrices corresponding to the weight $(1, \ldots, 1)$. The proof easily follows from Thm. 2.4 in [46].

Lemma 7.2 *Let* $\lambda \in \mathbb{N}^m$ *be a partition of* m. *Suppose* $D_\lambda = [D_{i,j}]$ *is an irreducible matrix representation of* GL_m *of type* λ *with respect to a basis of weight vectors. Then we have for all* $A \in \mathrm{GL}_m$ *that*

$$\mathrm{im}_\lambda(A) = \sum_i D_{i,i}(A) \ ,$$

where the sum is over all i *corresponding to basis vectors of weight* $(1, \ldots, 1)$.

If we extend the above sum over all indices, we get the character of D_λ evaluated at A. It is interesting, that this value can always be computed with a *polynomial number* of arithmetic operations in m. Compare Prop. 7.4 at the end of this section.

By combining our fast algorithm for evaluating representations of GL_m of the last chapter (Thm. 6.6) with the above lemma we get the following result.

Theorem 7.3 *Let* $\lambda \in \mathbb{N}^m$ *be a nonconstant partition of* m *and let* s_λ *denote the number of standard tableaus on the diagram of* λ. *One can compute* $\mathrm{im}_\lambda(A)$ *from* $A \in \mathrm{GL}_m$ *with a number of arithmetic operations bounded by*

$$O\big(m^2(\mathrm{mult}(\lambda) + \log m)\, s_\lambda d_\lambda\big) \ .$$

The nonscalar complexity is bounded by $O(m^2 s_\lambda d_\lambda)$.

Hartmann [47] proved the upper bound $m^{6(m-s)+4}$ for the nonscalar complexity to evaluate permanents corresponding to partitions with at most s parts. Barvinok [5] showed the upper bound $O(m^3 d_\lambda^4)$ for the total complexity. Our bound improves Barvinok's, as $s_\lambda \leq d_\lambda$ and $\mathrm{mult}(\lambda) \leq d_\lambda$.

To compare our bound with those of Hartmann, consider hook partitions $\lambda = (k, 1, \ldots, 1) \in \mathbb{N}^m$. For such λ one can show that

$$s_\lambda = \binom{m-1}{k-1}, \quad d_\lambda = \frac{m+k-1}{m} \binom{m+k-2}{m-k,\, k-1,\, k-1} \ ,$$

hence $m^2 s_\lambda d_\lambda \leq m^2 18^m$. This is considerably smaller than Hartmann's bound m^{6k} if $k \geq m/2$.

For permanents, our theorem yields the bound $O(m^{1.5} 4^m \log m)$, which is not too far away from the best known upper bound $O(m 2^m)$ due to Ryser [93].

We close this section by showing that the characters of GL_m can be evaluated very rapidly. Let $2 \leq \omega < 3$ denote the exponent of matrix multiplication (cf. [21]).

Proposition 7.4 *Let* $\epsilon > 0$. *Then for all* m, *the character* $\mathrm{Tr}(D_\lambda(A))$ *can be evaluated at a given matrix* $A \in \mathrm{GL}_m$ *with* $O(\max\{\lambda_1, m\}^{\omega + \epsilon})$ *arithmetic operations.*

Proof. The Schur polynomial of λ is defined as $S_\lambda = \text{Tr}(D_\lambda(\text{diag}(x_1, \ldots, x_m)))$. Let σ_i denote the ith elementary symmetric polynomial in m variables, and set $\sigma_i = 0$ if $i < 0$ or $i > m$. Moreover, let $\mu = (\mu_1, \ldots, \mu_{\lambda_1})$ be the partition conjugate to λ. Giambelli's formula states that (cf. [37, §A.1 (A.6)])

$$S_\lambda = \det[\sigma_{\mu_i + j - i}]_{1 \le i,j \le \lambda_1} \ .$$

Let $T^m + \sum_{i=1}^m (-1)^i c_i(A) T^{m-i}$ be the characteristic polynomial of the matrix A with eigenvalues $x_1, \ldots, x_m$. Then we have $c_i(A) = \sigma_i(x)$ for all i. Therefore, we get from Giambelli's formula that

$$\text{Tr}(D_\lambda(A)) = S_\lambda(x) = \det[c_{\mu_i + j - i}(A)]_{1 \le i,j \le \lambda_1} \ .$$

The algorithm is now as follows. First, compute the coefficients $c_i(A)$ of the characteristic polynomial for given A with $O(m^{\omega + \epsilon})$ arithmetic operations (cf. [21, §16.6]). Then compute $\text{Tr}(D_\lambda(A))$ by evaluating the determinant using $O(\lambda_1^{\omega + \epsilon})$ operations (cf. [21, §16.4]). $\qquad\qquad\square$

7.3   Completeness Results for Immanants

By the immanant polynomial IM_λ corresponding to a partition λ of n we understand the immanant im_λ evaluated at an n by n matrix of independent indeterminate entries. (Recall that we denote matrix functions with small letters, but the corresponding function evaluated at a matrix with independent indeterminate entries is written with capital letters.)

We have two completeness results for families of immanants over fields of characteristic zero. The first one extends a result by Hartmann [47].

Theorem 7.5 *Let (k_n) be a sequence of natural numbers with $n^\epsilon \le k_n \le n$ for some positive ϵ. Then the sequence of immanant polynomials corresponding to the hook partitions $(k_n, 1, \ldots, 1) \in \mathbb{N}^n$ is VNP-complete.*

Our second result answers a question posed by Strassen [106, Problem 14.2].

Theorem 7.6 *Let (s_m) be a polynomially bounded sequence of natural numbers. Then the sequence of immanant polynomials belonging to the rectangular partitions $(m, \ldots, m) \in \mathbb{N}^{m s_m}$ is VNP-complete.*

We will provide the proof of these theorems in the remaining sections of this chapter.

The complexity of immanants is still full of mysteries. For instance, we do not even know whether the sequence of immanant polynomials belonging to the partitions $(2, 2, \ldots, 2) \in \mathbb{N}^{2n}$ is VNP-complete!

Problem 7.1 What is the complexity of the immanant polynomials belonging to the partitions $(2, 2, \ldots, 2) \in \mathbb{N}^{2n}$?

We have the following conjecture.

Conjecture 7.1 Let $(\lambda^{(n)})$ be a sequence of partitions such that $|\lambda^{(n)}| = n$ and $\max \lambda^{(n)} \geq n^\epsilon$ for some positive ϵ. Then the corresponding sequence of immanant polynomials is VNP-complete.

We remark that our proofs also yield #P-completeness results for the problem to evaluate immanants at $0, 1$-matrices A. However, note that $\mathrm{im}_\chi(A)$ may be negative, with absolute value bounded by $n! \, n^n \leq n^{2n}$ for $A \in \{0,1\}^{n \times n}$. We will therefore interpret the evaluation problem below as the modified problem to compute $\mathrm{im}_\chi(A) + n^{2n}$ from A.

Corollary 7.7 *Assume in Theorems 7.5 and 7.6 that the maps $n \mapsto k_n$ and $n \mapsto s_n$ are polynomial time computable, when n is given in unary. Then the problem to evaluate the immanants corresponding to the hook partition $(k_n, 1, \ldots, 1) \in \mathbb{N}^n$ or to the rectangular partition $(n, \ldots, n) \in \mathbb{N}^{n s_n}$ at a given $0, 1$-matrix is #P-complete.*

7.4 Character Formulas for the Symmetric Group

We recall some character formulas for the symmetric group for later use. More information can be found in the books by Boerner [13], James and Kerber [51], or Fulton and Harris [37].

We will write $\lambda \vdash n$ in order to express that $\lambda = (\lambda_1, \ldots, \lambda_s)$ is a partition of n. As usual, we denote the character of S_n corresponding to λ by χ_λ. To a partition λ we may assign the strictly decreasing sequence

$$\ell = [\ell_1, \ldots, \ell_s] := (\lambda_1, \ldots, \lambda_s) + (s-1, s-2, \ldots, 1, 0)$$

in $\mathbb{N}^s$ which satisfies $\sum \ell_i = n + \binom{s}{2}$. (We use square brackets to distinguish ℓ notationally from a partition λ.) It is useful to index the irreducible characters of S_n by such sequences, thus we set $\chi_\ell := \chi_\lambda$. We can extend this definition to any $\ell \in \mathbb{N}^s$ satisfying $\sum \ell_i = n + \binom{s}{2}$ by requiring the function $\ell \mapsto \chi_\ell$ to be alternating. In particular, χ_ℓ vanishes if two components of ℓ are equal. We also include the case $n = 0$ by setting $\chi_{[s-1, \ldots, 0]}(1) := 1$.

Conjugacy classes of permutations in S_n are described by their *cycle format* $(\rho_1, \ldots, \rho_n)$, where ρ_i denotes the number of i-cycles. Clearly, we have $\sum_i i \rho_i = n$. It will be convenient to write cycle formats in frequency notation $\rho = 1^{\rho_1} \cdots n^{\rho_n}$, or shorter $\rho \models n$ in order to express that ρ is a cycle format of n. Moreover, we set $\chi_\ell(\rho) := \chi_\ell(\pi)$, where π is any permutation with cycle format ρ.

Let $u_{s,i} := Z_1^i + Z_2^i + \ldots Z_s^i$ denote the ith power sum in the indeterminates $Z_1, \ldots, Z_s$, and let

$$\Delta_s := \det(Z_i^{s-j})_{1 \leq i, j \leq s} = \prod_{i < j}(Z_i - Z_j)$$

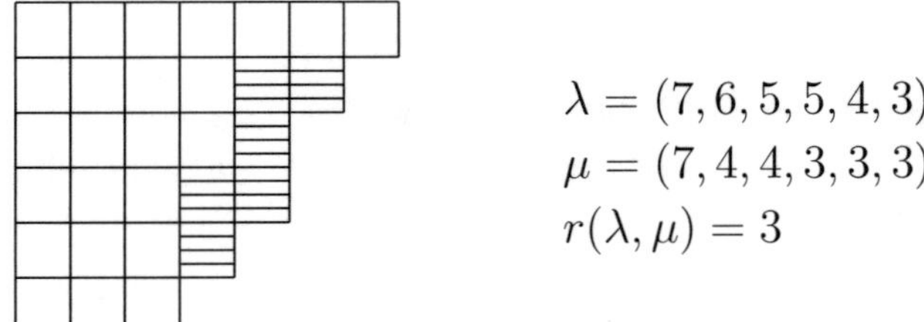

$$\lambda = (7,6,5,5,4,3)$$
$$\mu = (7,4,4,3,3,3)$$
$$r(\lambda,\mu) = 3$$

Figure 7.1: A skew hook for the Young diagram of λ.

be the discriminant. The characters of S_n are determined by the remarkable formula of Frobenius (cf. [37, 4.10, p. 49]):

$$(7.1) \qquad \Delta_s \cdot u_{s,1}^{\rho_1} \cdots u_{s,n}^{\rho_n} = \sum_{\ell} \chi_\ell(\rho) Z_1^{\ell_1} \cdots Z_s^{\ell_s} \ ,$$

where the sum is over all $\ell \in \mathbb{N}^s$ satisfying $\sum \ell_i = n + \binom{s}{2}$. From this formula one easily deduces Frobenius' recursion formula for the characters of S_n (cf. [13, VI §3]): let $1 \leq h \leq n$ and $\rho \models n - h$. Then we have for all $\ell \in \mathbb{N}^s$ satisfying $\sum \ell_i = n + \binom{s}{2}$

$$(7.2) \qquad \chi_\ell(\rho \cdot h) = \sum_i \chi_{[\ell_1,\ldots,\ell_{i-1},\ell_i-h,\ell_{i+1},\ldots,\ell_s]}(\rho) \ ,$$

where the sum is over all $1 \leq i \leq s$ such that $\ell_i \geq h$, and with the property that $\ell_1,\ldots,\ell_{i-1},\ell_i - h,\ell_{i+1},\ldots,\ell_s$ are pairwise distinct numbers.

Sometimes it is convenient to use a recursion formula related to (7.2), the so-called Murnaghan-Nakayama rule. We recall that a partition $\lambda \vdash n$ can be represented by its *(Young) diagram* $\{(i,j) \mid 1 \leq j \leq \lambda_i\}$, which should be visualized as a left-justified arrangement of λ_i boxes in the ith row. A *skew hook* for λ is a connected region of boundary boxes for its diagram such that removing them leaves a diagram for another partition μ. We denote by $r(\lambda,\mu)$ the number of vertical steps in the skew hook, i.e., one less than the number of rows in the hook (cf. Fig. 7.1). The Murnaghan-Nakayama rule now reads as follows: For $\lambda \vdash n$, $1 \leq h \leq n$, and $\rho \models n - h$ we have

$$(7.3) \qquad \chi_\lambda(\rho \cdot h) = \sum_\mu (-1)^{r(\lambda,\mu)} \chi_\mu(\rho) \ ,$$

where the sum is over all partitions $\mu \vdash n - h$ that can be obtained from λ by removing a skew hook containing h boxes. (Cf. [51, 2.4.7, p. 60] or [37, p. 59].)

As an example, let us compute the value of χ_λ on an n-cycle. The Murnaghan-Nakayama rule implies immediately that $\chi_\lambda(n^1) = (-1)^i$ if λ equals a hook partition $(n - i, 1, \ldots, 1)$, and that $\chi_\lambda(n^1) = 0$ otherwise.

For the rest of this chapter, we will denote the characters corresponding to the hook partition $(n - i, 1, \ldots, 1) \vdash n$ by $\chi_{n,i}$ and call them *hook characters*.

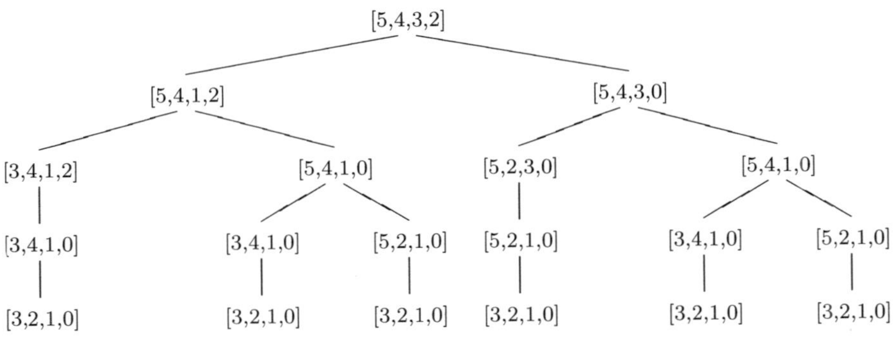

Figure 7.2: The tree $T_{2,2}$.

The corresponding immanant polynomials will be called *hook immanants* and denoted by $\mathrm{HI}_{n,i}$ for $0 \le i < n$. For instance $\mathrm{HI}_{n,0} = \mathrm{PER}_n$ and $\mathrm{HI}_{n,n-1} = \mathrm{DET}_n$.

Now assume $\rho \models n$, $\rho \ne n^1$. The orthogonality relations (cf. [37, I §2.2]) and the above observation about $\chi_\lambda(n^1)$ imply that

$$(7.4) \qquad \sum_{\lambda \vdash n} \chi_\lambda(n^1)\chi_\lambda(\rho) = \sum_{i=0}^{n-1}(-1)^i \chi_{n,i}(\rho) = 0 \ .$$

From this one easily concludes the following formula due to Merris [84]

$$(7.5) \qquad \sum_{i=0}^{n-1}(-1)^i \, \mathrm{HI}_{n,i} = n\mathrm{HC}_n \ .$$

We illustrate Frobenius' recursion formula (7.2) by computing the value of $\chi_{(2,2,2,2)}$ for permutations in S_8 of cycle format 2^4. To the partition $(2,2,2,2)$ there corresponds the sequence $\ell = [5,4,3,2]$. The recursive application of equation (7.2) can be illustrated by the tree $T_{2,2}$ in Fig. 7.2. $T_{2,2}$ has depth 4 and its nodes carry a label $\ell \in \mathbb{N}^4$ having distinct components. The meaning of this tree is the following: Consider a node with label ℓ at level $1 \le t \le 4$ and let $\ell^{(1)}, \ldots, \ell^{(M)}$ be the labels of the sons of this node. Then we have by (7.2) that

$$\chi_\ell(2^t) = \sum_{i=1}^{M} \chi_{\ell^{(i)}}(2^{t-1}) \ .$$

From this it follows that $\chi_{[5,4,3,2]}(2^4)$ is the sum of $\chi_{[3,2,1,0]}(1)$ over all leaves of the tree. Hence $\chi_{[5,4,3,2]}(2^4) = 6\chi_{[3,2,1,0]}(1) = 6$.

By abuse of notation, we will write χ_{m^s} for the character corresponding to a rectangular partition $(m, \ldots, m) \vdash sm$, and IM_{m^s} for the corresponding immanant polynomial.

The next lemma generalizes the above example and expresses certain values of χ_{m^s} by hook characters. The lemma will be crucial in our completeness proofs.

Lemma 7.8 *Let* $s = qm + r$, $1 \le r \le m$, $q \ge 0$. *Then we have for all cycle formats* $\rho \models m$, $\rho \ne m^1$, *that*

$$\chi_{m^s}(m^{s-1} \cdot \rho) = \gamma_{m,s} \sum_{i=0}^{r-1}(-1)^i \chi_{m,i}(\rho)$$

and

$$\chi_{m^s}(m^s) = r\gamma_{m,s} + m\beta_{m,s} \ ,$$

where

$$\gamma_{m,s} = \frac{(s-1)!}{q!^{m-r}(q+1)!^r}, \quad \beta_{m,s} = q\gamma_{m,s} \ .$$

Proof. As in the above example (Fig. 7.2) we can describe the recursive application of Frobenius' recursion formula (7.2) for computing $\chi_{m^s}(m^s)$ by a labeled tree $T_{m,s}$. This tree is built up as follows. The root carries the label

$$\ell^{(0)} := [m + s - 1, m + s - 2, \ldots, m] = (m, m, \ldots, m) + (s - 1, s - 2, \ldots, 0) \ .$$

To an already constructed node with label ℓ we create a son for every $\ell - me_k$ which has nonnegative and distinct components. Here $e_k \in \{0,1\}^s$ denotes the canonical basis vector having a 1 at position $1 \le k \le s$.

We will soon prove that all leaves of $T_{m,s}$ carry the label $[s - 1, \ldots, 1, 0]$. Using this already, we see that the nodes one level above the leaves carry a label of the form $L_k := [s - 1, \ldots, 1, 0] + me_k$ for $1 \le k \le m$. Let α_k denote the number of these nodes. By a repeated application of the recursion formula (7.2) we obtain for all $\rho \models m$

$$\chi_{m^s}(m^{s-1} \cdot \rho) = \sum_{k=1}^{m} \alpha_k \chi_{L_k}(\rho) \ .$$

A k-cycle transforms the sequence $(m-k+1, 1, \ldots, 1, 0, \ldots, 0) + (s-1, \ldots, 1, 0)$ into L_k, thus χ_{L_k} equals up to a sign a hook character: $\chi_{L_k} = (-1)^{k-1}\chi_{m,k-1}$. Now the point is that

$$(7.6) \qquad \alpha_1 = \ldots = \alpha_r, \ \alpha_{r+1} = \ldots = \alpha_m \ .$$

Assuming this for the moment, we may conclude for $\rho \models m$, $\rho \ne m^1$, that

$$\begin{aligned}
\chi_{m^s}(m^{s-1} \cdot \rho) &= \alpha_1 \sum_{i=0}^{r-1}(-1)^i \chi_{m,i}(\rho) + \alpha_{r+1} \sum_{i=r}^{m-1}(-1)^i \chi_{m,i}(\rho) \\
&= (\alpha_1 - \alpha_{r+1}) \sum_{i=0}^{r-1}(-1)^i \chi_{m,i}(\rho) \ ,
\end{aligned}$$

where we have used formula (7.5) in the last equality. We also get (recall that $\chi_{m,i}(m^1) = (-1)^i$)

$$\chi_{m^s}(m^s) = r\alpha_1 + (m-r)\alpha_{r+1} = r(\alpha_1 - \alpha_{r+1}) + m\alpha_{r+1} \ .$$

So it remains to show (7.6) and to check that indeed $\alpha_1 - \alpha_{r+1} = \gamma_{m,s}$ and $\alpha_{r+1} = q\gamma_{m,s}$.

To a leaf of $T_{m,s}$ there corresponds bijectively the path from the root to this leaf, which can be uniquely described by the sequence of labels $(\ell^{(0)}, \ldots, \ell^{(s)})$ of the nodes along this path. By assigning to $\ell^{(i)}$ the set $A_i \subseteq B := \{0, 1, \ldots, m+s-1\}$ of its components, we get a sequence $(A_0, \ldots, A_s)$ of subsets of B all having cardinality s, and which satisfy

$$(7.7) \qquad\qquad A_{i+1} = (A_i \setminus \{a_i\}) \cup \{a_i - m\}$$

with some $a_i \in A_i$ such that $a_i \geq m$ and $a_i - m \notin A_i$. It is easy to see that this correspondence is in fact a bijection between the paths of $T_{m,s}$ from the root to a leaf and the sequences $(A_0, \ldots, A_s)$ of subsets of B satisfying (7.7) and such that $A_0 = \{m, m+1, \ldots, m+s-1\}$.

Now consider the complements $\overline{A}_i := B \setminus A_i$. They are all of cardinality m. By induction on i one shows that the remainders modulo m of the elements of $\overline{A}_i$ are pairwise distinct. Hence we may write

$$(7.8) \qquad\qquad \overline{A}_i = \{p_1^{(i)}m, p_2^{(i)}m + 1, \ldots, p_m^{(i)}m + m - 1\}$$

with a uniquely determined vector $p^{(i)} = (p_1^{(i)}, \ldots, p_m^{(i)})$ contained in

$$W := \{0, 1, \ldots, q+1\}^r \times \{0, 1, \ldots, q\}^{m-r} \ .$$

Let $a_i \in A_i$ be as in (7.7). As $a_i - m \notin A_i$, we have $a_i - m = p_{\mu_i}^{(i)}m + \mu_i - 1$ for some $1 \leq \mu_i \leq m$. On the other hand, $a_i = (p_{\mu_i}^{(i)} + 1)m + \mu_i - 1$ is by construction not contained in A_{i+1}. This implies that

$$(7.9) \qquad\qquad p^{(i+1)} = p^{(i)} + e'_{\mu_i} \ ,$$

where $e'_{\mu_i} \in \{0, 1\}^m$ is the canonical basis vector having a 1 at position μ_i. We can thus regard $(p^{(0)}, \ldots, p^{(s)})$ as a *walk* in W which starts in $p^{(0)} = \underline{0} := (0, \ldots, 0)$ and must end in the opposite corner $p^{(s)} = w$, where $w := (q+1, \ldots, q+1, q, \ldots, q)$. In this walk, a successor of a point is obtained by incrementing exactly one coordinate by 1.

It is now straightforward to check that we have found a bijection between the leaves of $T_{m,s}$ and the above described walks in W from $\underline{0}$ to the opposite corner w. In particular, all leaves of $T_{m,s}$ carry the same label $[s-1, \ldots, 1, 0]$ corresponding to w, as was claimed at the beginning of the proof.

In the sequel we will assume that $q \geq 1$. (The case $q = 0$ can be checked separately.) A node N of $T_{m,s}$ one level above the leaves corresponds to a walk in W ending in one of the points $w - e'_M$, where $1 \leq M \leq m$. To such a walk in turn there corresponds a sequence of sets $(A_0, \ldots, A_{s-1})$. Assume first that $1 \leq M \leq r$. Then it is easily checked that

$$A_{s-1} = (\{0, 1, \ldots, s-1\} \setminus \{qm + M - 1\}) \cup \{(q+1)m + M - 1\} \ .$$

By comparing this with the set of components of L_k, we obtain $qm + M - 1 = s - k$, hence $k = r - M + 1$. We conclude that the node N carries the label L_{r-M+1}. The number of nodes of level $s-1$ carrying the label L_{r-M+1} equals the number of walks in W from $\underline{0}$ to $w - e'_M$. Such a walk is uniquely described (cf. (7.9)) by a sequence $(M_0, M_1, \ldots, M_{s-2})$ in $\{1, 2, \ldots, m\}^{s-1}$, in which $\mu \in \{M\} \cup \{r+1, \ldots, m\}$ occurs with frequency q, and $\mu \in \{1, \ldots, r\} \setminus \{M\}$ occurs with frequency $q + 1$. The number of these sequences equals the multinomial coefficient

$$\alpha := \frac{(s-1)!}{q!^{m-r+1}(q+1)!^{r-1}} \ .$$

This proves that $\alpha_1 = \ldots = \alpha_r = \alpha$.

In the case $r < M \leq m$ one can show similarly that N carries the label $L_{r-M+m+1}$ and that the number of nodes carrying this label equals

$$\beta := \frac{(s-1)!}{(q-1)! \, q!^{m-r-1}(q+1)!^r} \ ,$$

which yields $\alpha_{r+1} = \ldots = \alpha_m = \beta$. A straightforward calculation shows that indeed $\alpha - \beta = \gamma_{m,s}$ and $\beta = q\gamma_{m,s}$, where $\gamma_{m,s} = (s-1)!/(q!^{m-r}(q+1)!^r)$. $\square$

7.5 p-Definability of Immanants

Frobenius formula (7.1) for the generating function of S_n-characters allows to express immanants as coefficients of p-computable families of polynomials. Based on this observation and some of the closure properties of VNP discussed in Chap. 2, we can prove the following proposition.

Proposition 7.9 *The sequence of immanants* $(\mathrm{IM}_{\lambda(n)})$ *is p-definable over* $\mathbb{Q}$ *for any sequence of partitions* $(\lambda^{(n)})$, *where* $\lambda^{(n)} \vdash n$.

Proof. Recall the cycle format polynomials CF_ρ from Sect. 3.3.3. They are defined as

$$\mathrm{CF}_\rho := \sum_\pi \prod_{i=1}^n X_{i,\pi(i)} \ ,$$

where the sum is over all permutations π having cycle format $\rho \models n$.

We multiply Frobenius' formula (7.1) for $s = n$ with CF_ρ and take the sum over all cycle formats $\rho \models n$. This yields

$$F_n := \Delta_n \sum_{\rho \models n} u_{n,1}^{\rho_1} \cdots u_{n,n}^{\rho_n} \, \mathrm{CF}_\rho = \sum_\ell \Big(\sum_{\rho \models n} \chi_\ell(\rho) \, \mathrm{CF}_\rho \Big) Z_1^{\ell_1} \cdots Z_n^{\ell_n}.$$

The immanant $\mathrm{IM}_\lambda = \sum_\rho \chi\ell(\rho)\mathrm{CF}_\rho$ appears in F_n as the coefficient of the power product $\prod_j Z_j^{\ell_j} = \prod_j Z_j^{\lambda_j + n - j}$. By Thm. 2.19(3) it is therefore sufficient to prove that (F_n) is p-definable. Let T_{ij} be further indeterminates for $1 \le i, j \le n$ and define

$$G_n := \sum_{\rho \models n} \mathrm{CF}_\rho \prod_{(i,j):j \le \rho_i} T_{i,j}.$$

By substituting $T_{i,j}$ by $u_{n,i} = Z_1^i + \ldots + Z_n^i$ in G_n and multiplying the resulting polynomial with $\Delta_n = \prod_{i<j}(Z_i - Z_j)$, we obtain F_n. By Thm. 2.19(1),(2) it is therefore enough to show that the family (G_n) is p-definable. If we denote by $\rho_i(\pi)$ the number of i-cycles of π, we may write

$$
\begin{aligned}
G_n &= \sum_{\pi \in S_n} \Big(\prod_{\alpha \le n} X_{\alpha,\pi(\alpha)} \Big) \Big(\prod_{(i,j):j \le \rho_i(\pi)} T_{i,j} \Big) \\
&= \sum_{e,\epsilon \in \{0,1\}^{n \times n}} \phi_n(e,\epsilon) \prod_{1 \le \alpha,\beta \le n} X_{\alpha,\beta}^{e_{\alpha,\beta}} \prod_{1 \le i,j \le n} T_{i,j}^{\epsilon_{i,j}}
\end{aligned}
$$

with a uniquely determined function

$$\phi_n \colon \{0,1\}^{n \times n} \times \{0,1\}^{n \times n} \to \{0,1\}.$$

It is obvious that the extension of all ϕ_n to a function $\phi \colon \{0,1\}^* \to \{0,1\}$ is computable by a polynomial time Turing machine. Therefore, Valiant's criterion 2.20 implies that (G_n) is p-definable. $\qquad\square$

We can alternatively deduce Prop. 7.9 from the following interesting result due to Hepler [50], which shows that computing characters of the symmetric groups is a #P-complete problem. This result should be contrasted with Prop. 7.4, which implies that the evaluation of characters of GL_m is possible in polynomial time.

Theorem 7.10 (Hepler) *The function which assigns to $\lambda \vdash n$, $\rho \models n$ the value $\chi_\lambda(\rho) + n^n$ is #P-complete (n given in unary).*

In fact, to deduce Prop. 7.9, we only need the easy part of this theorem stating that the above function is contained in #P. Let a sequence of partitions $(\lambda^{(n)})$ be given, where $|\lambda^{(n)}| = n$. We can interpreted $\lambda^{(n)}$ as an "advice" for n in the sense of Karp and Lipton [61]. By Thm. 7.10, the function which assigns to the (description π of a) power product $\prod_{i=1}^n X_{i,\pi(i)}$ its coefficient in the polynomial $\mathrm{IM}_{\lambda^{(n)}} + n^n \mathrm{PER}_n$ is in the class #P/poly. Therefore, by Valiant's criterion 2.20, the family $(\mathrm{IM}_{\lambda^{(n)}} + n^n \mathrm{PER}_n)_n$ is p-definable. This again implies Prop. 7.9.

7.6   Completeness Proofs

We provide now the proofs of our completeness results for immanants as stated in Sect. 7.3. Let us recall the correspondence between square matrices and (edge-) weighted digraphs introduced in Sect. 2.2.2. We can regard an n by n matrix $A = [a_{i,j}]$ over some field as the adjacency matrix of a weighted digraph G on n nodes, where $a_{i,j} \neq 0$ gives the weight of the edge from node i to node j, and there is no such edge if $a_{i,j} = 0$. We define the weight $\mathrm{wt}(\mathcal{C})$ of a subgraph $\mathcal{C}$ of G as the product of the weight of all its edges. The permutations $\pi \in S_n$ with $\prod_{i=1}^{n} a_{i,\pi(i)} \neq 0$ correspond bijectively to the cycle covers $\mathcal{C}$ of G, i.e., n-node subgraphs of G consisting of node-disjoint cycles. Note that $\mathrm{wt}(\mathcal{C}) = \prod_i a_{i,\pi(i)}$ if $\mathcal{C}$ corresponds to π. A cycle cover $\mathcal{C}$ has a (cycle) format $\rho \models n$, where ρ_i counts the number of i-cycles of $\mathcal{C}$. We define now the *cycle format value* $\mathrm{cf}_\rho(G)$ of a weighted digraph G as the sum of the weights of all cycle covers of G having the format ρ. A special case of this is the *Hamilton cycle value* $\mathrm{hc}(G) := \mathrm{cf}_n(G)$. The *immanant* $\mathrm{im}_\lambda(G)$ *of* G is defined as

$$\mathrm{im}_\lambda(G) := \sum_{\rho \models n} \chi_\lambda(\rho)\mathrm{cf}_\rho(G) \ .$$

If all the weights of G are either rationals or indeterminates, then $\mathrm{im}_\lambda(G)$ is obviously a projection of IM_λ. Moreover, if DK_n denotes the complete weighted digraph DK_n on n nodes with edge (i,j) carrying the indeterminate weight $X_{i,j}$, then $\mathrm{im}_\lambda(\mathrm{DK}_n) = \mathrm{IM}_\lambda$. According to our convention, we will denote the value of the hook immanant $\mathrm{HI}_{n,i}$ on G by $\mathrm{hi}_{n,i}(G)$. The cycle format polynomial satisfies $\mathrm{CF}_\rho = \mathrm{cf}_\rho(\mathrm{DK}_n)$.

We explain now our strategy for proving completeness for a family of immanants. Let $1 \leq h \leq n$ and G be the disjoint union of the complete weighted digraph DK_{n-h} on $n - h$ nodes with a directed cycle Z of length h, all of whose edges carry the weight 1. Every cycle cover $\mathcal{C}$ of G consists of a cycle cover $\mathcal{C}'$ of DK_{n-h} and the h-cycle Z, and we have $\mathrm{wt}(\mathcal{C}) = \mathrm{wt}(\mathcal{C}')$. Therefore, we obtain for $\rho \models n - h$

$$\mathrm{cf}_{\rho \cdot h^1}(G) = \mathrm{cf}_\rho(DK_{n-h}) = \mathrm{CF}_\rho \ ,$$

and all other cycle format values of G vanish. From this and the Murnaghan-Nakayama rule (7.3) we obtain

$$\begin{aligned}
\mathrm{im}_\lambda(G) &= \sum_{\rho \models n-h} \chi_\lambda(\rho \cdot h^1)\, \mathrm{CF}_\rho = \sum_{\rho \models n-h} \sum_\mu (-1)^{r(\lambda,\mu)} \chi_\mu(\rho)\, \mathrm{CF}_\rho \\
&= \sum_\mu (-1)^{r(\lambda,\mu)} \sum_{\rho \models n-h} \chi_\mu(\rho)\, \mathrm{CF}_\rho = \sum_\mu (-1)^{r(\lambda,\mu)} \mathrm{IM}_\mu \ .
\end{aligned}$$

Let us summarize this important insight in slightly more general form.

Lemma 7.11 *Let* $\lambda^{(1)}, \ldots, \lambda^{(t)} \vdash n$, $\alpha_1, \ldots, \alpha_t \in \mathbb{Q}$, *and* $1 \leq h \leq n$. *Then the linear combination of immanants* $\sum_{i=1}^{t} \alpha_i \mathrm{IM}_{\lambda^{(i)}}$ *has as a projection the linear combination of immanants*

$$\sum_{i=1}^{t} \alpha_i \sum_{\mu^{(i)}} (-1)^{r(\lambda^{(i)}, \mu^{(i)})} \mathrm{IM}_{\mu^{(i)}} \ ,$$

where $\mu^{(i)}$ *runs over all partitions* $\mu^{(i)} \vdash n - h$ *which can be obtained from* $\lambda^{(i)}$ *by removing a skew-hook containing* h *boxes.*

We begin now with the proof of Thm. 7.6. Let (s_m) be a p-bounded sequence of natural numbers and put $s_m = q_m m + r_m$ with $1 \leq r_m \leq m$, $q_m \geq 0$. By Prop. 7.9 the family $(\mathrm{IM}_{m^{s_m}})$ of immanant polynomials is p-definable, so it suffices to show that this family has a complete family as a p-projection.

Let G_m be the disjoint union of the complete weighted digraph DK_m with $s_m - 1$ directed cycles of length m, all of whose edges having weight 1. Then we obtain as before that

$$f_m := \mathrm{im}_{m^{s_m}}(G) = \sum_{\rho \models m} \chi_{m^{s_m}}(m^{s_m - 1} \cdot \rho) \, \mathrm{CF}_\rho$$

is a projection of $\mathrm{IM}_{m^{s_m}}$. Lemma 7.8 implies that

$$f_m = \sum_{\substack{\rho \models m \\ \rho \neq m}} \gamma_{m, s_m} \sum_{i=0}^{r_m - 1} (-1)^i \chi_{m,i}(\rho) \, \mathrm{CF}_\rho + (r_m \gamma_{m, s_m} + m \beta_{m, s_m}) \mathrm{CF}_m$$

$$= \sum_{\rho \models m} \gamma_{m, s_m} \sum_{i=0}^{r_m - 1} (-1)^i \chi_{m,i}(\rho) \, \mathrm{CF}_\rho + m \beta_{m, s_m} \mathrm{CF}_m \ .$$

Therefore, we obtain

$$(7.10) \qquad f_m = \gamma_{m, s_m} \sum_{i=0}^{r_m - 1} (-1)^i \, \mathrm{HI}_{m,i} + m \beta_{m, s_m} \mathrm{HC}_m \ .$$

In the special case $r_m = m$ we conclude with formula (7.5) that f_m is a nonzero scalar multiple of HC_m. This shows that HC is a p-projection of $\mathrm{IM}_{m^{s_m}}$ and proves Thm. 7.6 in the situation of square diagrams!

The general case of rectangular diagrams is considerably more complicated. We need the following technical lemmas, the first of which is proved similarly as Thm. 2 in Hartmann [47].

Lemma 7.12 (1) *Let us assume that* $0 \leq r < n$ *and* $\alpha_0, \ldots, \alpha_r, \beta \in \mathbb{Q}$. *Put* $p := \lfloor n/(r+1) \rfloor$ *and* $\gamma := \sum_{i=0}^{r} (-1)^i \alpha_i$. *Then* $\gamma \mathrm{PER}_p + \beta \mathrm{HC}_p$ *is a projection of* $\sum_{i=0}^{r} \alpha_i \mathrm{HI}_{n,i} + \beta \mathrm{HC}_n$.

(2) *Let $1 \le r < n$, $\alpha_0, \alpha_r, \ldots, \alpha_{n-1}, \beta \in \mathbb{Q}$. Put*

$$p := \lfloor \frac{n}{n-r+\delta} \rfloor, \quad R := n - p(n-r+\delta), \quad \gamma := (-1)^{n+R-1} \sum_{i=r}^{n-1} (-1)^i \alpha_i \; ,$$

where $\delta \in \{0,1\}$, $\delta \equiv n-r+1 \bmod 2$. Then $\alpha_0 \mathrm{PER}_p + \gamma \mathrm{DET}_p + \beta \mathrm{HC}_p$ is a projection of $\alpha_0 \mathrm{PER}_n + \sum_{i=r}^{n-1} \alpha_i \mathrm{HI}_{n,i} + \beta \mathrm{HC}_n$.

Proof. We note first that by the Murnaghan-Nakayama rule we have for the hook characters $\chi_{n,i}$ that $\chi_{n,i}(\rho) = (-1)^i$ if the cycle format ρ does not contain cycles of length at most i.

(1) For each $i \in \{2, \ldots, p\}$ we introduce a cycle of length $r+1$ with edge weights as indicated in the figure.

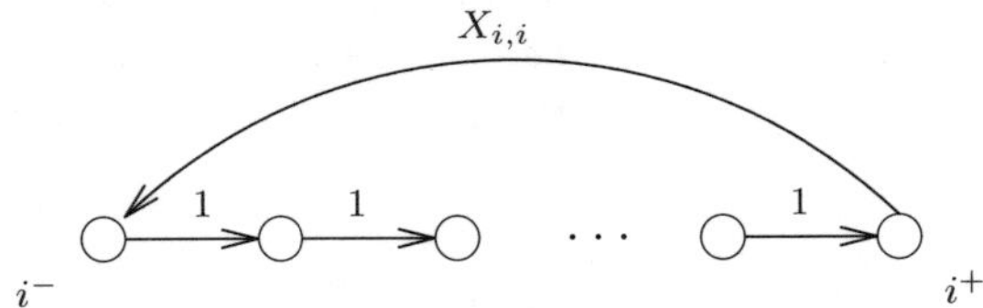

Analogously, we introduce a cycle of length $n - (p-1)(r+1)$ for $i = 1$. Moreover, we connect the node i^+ with the node j^- by a (directed) edge of weight $X_{i,j}$, for all distinct $i, j \in \{1, 2, \ldots, p\}$. The resulting weighted digraph G has n nodes and directed girth $\ge r+1$, that is, all cycles of G have length at least $r+1$. A cycle cover $\tilde{\mathcal{C}}$ of G corresponds bijectively to a cycle cover $\mathcal{C}$ of the complete weighted digraph DK_p. Moreover, $\mathrm{wt}(\tilde{\mathcal{C}}) = \mathrm{wt}(\mathcal{C})$, and $\tilde{\mathcal{C}}$ is a Hamilton cycle iff $\mathcal{C}$ is so.

By the observation at the beginning of the proof we have $\chi_{n,i}(\tilde{\mathcal{C}}) = (-1)^i$ for all $0 \le i \le r$ and cycle covers $\tilde{\mathcal{C}}$ of G (which we identify with permutations in S_n). We get therefore for $0 \le i \le r$

$$\mathrm{hi}_{n,i}(G) = \sum_{\mathcal{C}} \chi_{n,i}(\tilde{\mathcal{C}}) \mathrm{wt}(\tilde{\mathcal{C}}) = (-1)^i \sum_{\mathcal{C}} \mathrm{wt}(\mathcal{C}) = (-1)^i \, \mathrm{PER}_p \; ,$$

and $\mathrm{hc}(G) = \mathrm{HC}_p$, hence

$$\left(\sum_{i=0}^{r} \alpha_i \mathrm{hi}_{n,i} + \beta \mathrm{hc} \right)(G) = \gamma \mathrm{PER}_p + \beta \mathrm{HC}_p \; .$$

This proves statement (1).

(2) We assume that $n-r$ is odd, thus $\delta = 0$. (In the case where $n-r$ is even one can argue analogously.) As in the proof of (1) we construct a weighted digraph G by introducing for each $i \in \{2, \ldots, p\}$ a cycle of length $n - r$, and for $i = 1$ a cycle of length $n - (p-1)(n-r) = n-r+R$. Again, a cycle cover $\tilde{\mathcal{C}}$ of G corresponds bijectively to a cycle cover $\mathcal{C}$ of DK_p. Note that an ℓ-cycle

of DK_p which does not pass through 1 is assigned to a cycle of G having length $\ell(n-r)$, and that $\ell \equiv \ell(n-r)$ mod 2. To an ℓ-cycle of DK_p passing through 1 there corresponds a cycle of DK_p having length $\ell(n-r)+R$, which is congruent to $\ell + R$ modulo 2. From this we see that for any cycle cover $\mathcal{C}$ of DK_p

$$\mathrm{sgn}(\tilde{\mathcal{C}}) = (-1)^R \mathrm{sgn}(\mathcal{C}) \ .$$

By interchanging rows and columns in the hook diagram $(n-i,1,\ldots,1)$ we get the diagram of $(i+1,1,\ldots,1)$. Therefore, the corresponding characters are conjugated:

$$\chi_{n,i}(\tilde{\mathcal{C}}) = \mathrm{sgn}(\tilde{\mathcal{C}})\chi_{n,n-i-1}(\tilde{\mathcal{C}})$$

(cf. [37, Ex. 4.4, p. 47]). On the other hand, we have for all $r \le i < n$ and cycle covers $\tilde{\mathcal{C}}$ that $\chi_{n,n-i-1}(\tilde{\mathcal{C}}) = (-1)^{n-i-1}$, since $\mathrm{girth}(G) \ge n-r$. From these observations we conclude that for all $r \le i < n$

$$\begin{aligned}
\mathrm{hi}_{n,i}(G) &= \sum_{\mathcal{C}} \chi_{n,i}(\tilde{\mathcal{C}})\mathrm{wt}(\tilde{\mathcal{C}}) \\
&= \sum_{\mathcal{C}} (-1)^R \mathrm{sgn}(\mathcal{C})\chi_{n,n-i-1}(\tilde{\mathcal{C}})\mathrm{wt}(\mathcal{C}) \\
&= (-1)^{n+R-1}(-1)^i \sum_{\mathcal{C}} \mathrm{sgn}(\mathcal{C})\mathrm{wt}(\mathcal{C}) \\
&= (-1)^{n+R-1}(-1)^i \, \mathrm{DET}_p \ .
\end{aligned}$$

Taking into account that $\mathrm{hc}(G) = \mathrm{HC}_p$ and $\mathrm{per}(G) = \mathrm{PER}_p$, the claim follows now readily. $\qquad\square$

Lemma 7.13 *Let* $\alpha, \beta \in \mathbb{Q}$, $\alpha \ne 0$. *Then:*

(1) PER_{n-1} *is a projection of* $\alpha\mathrm{PER}_n + \beta\mathrm{DET}_n$.

(2) HC_{n-2} *is a projection of* $\alpha\mathrm{HC}_n + \beta\mathrm{DET}_n$.

Proof. (1) We concatenate the matrix $[X_{i,j}]_{1 \le i,j < n}$ of indeterminates with the column $[0,\ldots,0,(2\alpha)^{-1}]^T$ and repeat in the resulting matrix the last row. In this way we get an n by n matrix M having determinant zero. By expanding along the last column we get $\mathrm{per}(M) = 2(2\alpha)^{-1}\mathrm{per}([X_{i,j}]_{i,j<n}) = \alpha^{-1}\mathrm{PER}_{n-1}$. Hence $(\alpha\,\mathrm{per} + \beta\,\mathrm{det})(M) = \mathrm{PER}_{n-1}$.

(2) Consider the n by n matrix

$$M := \begin{bmatrix}
0 & 0 & 1 & X_{1,2} & \cdots & X_{1,n} \\
\alpha^{-1} & 0 & 0 & 0 & \cdots & 0 \\
0 & 0 & 1 & X_{1,2} & \cdots & X_{1,n} \\
0 & X_{2,1} & 0 & X_{2,2} & \cdots & X_{2,n} \\
\vdots & \vdots & \vdots & \vdots & & \vdots \\
0 & X_{n,1} & 0 & X_{n,2} & \cdots & X_{n,n}
\end{bmatrix} \ .$$

As the first and third row of M coincide, we have $\det(M) = 0$. We claim that $\mathrm{hc}(M) = \alpha^{-1}\mathrm{HC}_{n-2}$. This is best seen by considering the weighted digraph with adjacency matrix M which is built up as follows from DK_{n-2}. Assume that $\{1, 2, \ldots, n-2\}$ is the set of nodes of DK_{n-2}. We delete the loop $(1,1)$ and split the node 1 into nodes 1^- and 1^+. The edges $(i,1)$ of DK_{n-2} are thus replaced by edges $(i, 1^-)$, and the edges $(1, i)$ of DK_{n-2} are replaced by edges $(1^+, i)$, for $2 \le i \le n$. Now we introduce a new node O, an edge $(1^-, O)$ of weight α^{-1}, an edge $(O, 1^+)$ of weight 1, a loop $(1^+, 1^+)$ of weight 1, and edges (O, i) of weight $X_{1,i}$, for $2 \le i \le n$. The reader may easily verify that the resulting weighted digraph G has indeed the adjacency matrix M. On the other hand, it is clear from the construction that every Hamilton cycle of G passes through the edges $(1^-, O)$ and $(O, 1^+)$. This shows that indeed $\mathrm{hc}(G) = \alpha^{-1}\mathrm{hc}(\mathrm{DK}_{n-2}) = \alpha^{-1}\mathrm{HC}_{n-2}$. Altogether, we have $(\alpha\,\mathrm{hc} + \beta\,\det)(G) = \mathrm{HC}_{n-2}$ which proves the claim. $\qquad\square$

We recall that a p-family (f_n) is said to be monotone iff f_n is a projection of f_{n+1} for all n. The above lemma shows in particular that PER is monotone. Also HC is monotone, which can be demonstrated similarly as part (b) of the above lemma. We will write $\varphi(n) \gtrsim \psi(n)$ for functions $\varphi, \psi \colon \mathbb{N} \to (0, \infty)$ iff $\liminf_{n \to \infty} \varphi(n)/\psi(n) \ge 1$.

We can now finally provide the proofs of Thm. 7.5 and Thm. 7.6.

Proof. (of Thm. 7.5) Let $\epsilon > 0$ and (i_n) be a sequence of natural numbers satisfying $n - i_n \ge n^\epsilon$. We have to prove that (HI_{n,i_n}) is VNP-complete. As we already know from Prop. 7.9 that (HI_{n,i_n}) is p-definable, it is sufficient to prove that PER is a p-projection of this p-family. We distinguish several cases.

CASE 1 $i_n \le \sqrt{n}$.
Let $p_n := \lfloor \frac{n}{i_n + 1} \rfloor$. Then $p_n \gtrsim \sqrt{n}$ and Lemma 7.12(1) implies that $(-1)^{i_n}\mathrm{PER}_{p_n}$ is a projection of HI_{n,i_n}.

CASE 2 $i_n > \sqrt{n}$.
Put $m := n - i_n - 1$ and let $m = q i_n + \overline{m}$, where $0 \le \overline{m} < i_n$. By applying Lemma 7.11 with $h = i_n$ we obtain that $f_1 := \mathrm{HI}_{n-i_n, i_n} + (-1)^{i_n - 1}\mathrm{HI}_{n-i_n, 0}$ is a projection of HI_{n,i_n} (cf. Fig. 7.3). Again applying this lemma with $h = i_n$ yields that

$$f_2 := \mathrm{HI}_{n-2i_n, i_n} + (-1)^{i_n - 1}\mathrm{HI}_{n-2i_n, 0} + (-1)^{i_n - 1}\mathrm{HI}_{n-2i_n, 0}$$

is a projection of f_1. If we continue in this way, we see that with $\overline{n} := n - q i_n = \overline{m} + i_n + 1$ the polynomial

$$f_q := \mathrm{HI}_{\overline{n}, i_n} + q(-1)^{i_n - 1}\mathrm{PER}_{\overline{n}}$$

is a projection of HI_{n,i_n}.

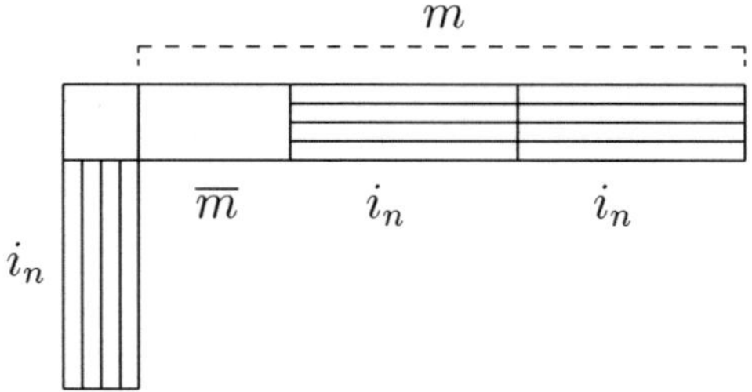

Figure 7.3: The hook diagram $(n - i_n, 1, \ldots, 1)$ when $q = 2$.

SUBCASE 2.1 $\overline{m} \geq n^{1/4}$.
We apply Lemma 7.11 with $h = i_n$ to f_q and get that $(q+1)(-1)^{i_n-1}\mathrm{PER}_{\overline{m}+1}$ is a projection of HI_{n,i_n} (recall that $\overline{m} < i_n$).

SUBCASE 2.2 $\overline{m} < n^{1/4}$ and $q \geq 1$.
We apply Lemma 7.12(2) to f_q and obtain

$$g_n := \gamma_n \mathrm{DET}_{p_n} + q(-1)^{i_n-1}\mathrm{PER}_{p_n}$$

as a projection of f_q for some $\gamma_n \in \{-1, 1\}$ and some

$$p_n \geq \lfloor \frac{n}{n - i_n + 1} \rfloor \geq \lfloor \frac{i_n}{\overline{m} + 2} \rfloor \gtrsim n^{1/4} \ .$$

By invoking Lemma 7.13(1), we see that $\mathrm{PER}_{p_n - 1}$ is a projection of g_n and thus of HI_{n,i_n}.

SUBCASE 2.3 $q = 0$.
As in Subcase 2.1 we apply Lemma 7.11 with $h = i_n$ to $f_q = \mathrm{HI}_{n,i_n}$ and get that $(-1)^{i_n-1}\mathrm{PER}_{\overline{m}+1}$ is a projection of HI_{n,i_n}. But $\overline{m} + 1 = n - i_n \geq n^\epsilon$ by assumption.

To summarize, let $\delta := \min\{\epsilon, 1/4\}$. We have shown the existence of a sequence $N_n \gtrsim n^\delta$ of natural numbers such that PER_{N_n} is a projection of HI_{n,i_n} for all sufficiently large n. Taking into account that PER is monotone, this implies that PER is a p-projection of $(\mathrm{HI}_{n,i_n})_n$ and finishes the proof. $\square$

Proof. (of Thm. 7.6) Let us go back to equation (7.10) and recall that f_m is a projection of the immanant polynomial $\mathrm{IM}_{m^s m}$ corresponding to a rectangular diagram. We distinguish two cases.

CASE 1 $1 \leq r_m \leq m - \sqrt{m}$.
We apply Lemma 7.11 with $h = 1$ and obtain that

$$g_m := \gamma_{m,s_m} \left(\mathrm{HI}_{m-1,0} + \sum_{i=1}^{r_m-1} (-1)^i \left[\mathrm{HI}_{m-1,i} + \mathrm{HI}_{m-1,i-1} \right] \right)$$

is a projection of $\gamma_{m,s_m} \sum_{i=0}^{r_m-1} (-1)^i \, \mathrm{HI}_{m,i}$. Recall that this is shown by adding to DK_{m-1} an isolated vertex with a loop (of weight 1). The resulting digraph does not have a Hamilton cycle. From this one easily sees that g_m is also a projection of f_m. The formula for g_m simplifies to (telescoping sum)

$$g_m = \gamma_{m,s_m} (-1)^{r_m-1} \mathrm{HI}_{m-1,r_m-1} \ .$$

If $r_m > m - \sqrt{m}$, then we define $g_m := \mathrm{HI}_{m-1,0}$. The family (g_m) is complete by Thm. 7.5.

CASE 2 $m - \sqrt{m} < r_m \leq m$.
By using relation (7.5) we can rewrite f_m as follows

$$\begin{aligned}
f_m &= \gamma_{m,s_m} \left(m\mathrm{HC}_m - \sum_{i=r_m}^{m-1} (-1)^i \, \mathrm{HI}_{m,i} \right) + m\beta_{m,s_m} \mathrm{HC}_m \\
&= \gamma_{m,s_m} \sum_{i=r_m}^{m-1} (-1)^{i+1} \mathrm{HI}_{m,i} + \kappa_m \mathrm{HC}_m \ ,
\end{aligned}$$

where $\kappa_m := m(\beta_{m,s_m} + \gamma_{m,s_m}) > 0$. Lemma 7.12(2) shows that

$$\varphi_n := c_m \mathrm{DET}_{p_m} + \kappa_m \mathrm{HC}_{p_m}$$

is a projection of f_m for some $c_m \in \mathbb{Q}$ and some

$$p_m \geq \lfloor \frac{m}{m - r_m + 1} \rfloor \geq \lfloor \frac{m}{\sqrt{m} + 1} \rfloor \geq \lfloor \sqrt{m} \rfloor - 1 \ .$$

Lemma 7.13(2) together with the fact that HC is monotone imply that the Hamilton cycle polynomial $h_m := \mathrm{HC}_{\lfloor \sqrt{m} \rfloor - 3}$ is a projection of φ_m and thus of f_m. The family (h_m) is monotone and complete, as HC has these properties.

    To summarize, we have proved that some mixture of the families (g_m) and (h_m) is a p-projection of (f_m), and thus of $(\mathrm{IM}_{m^{s_m}})$. Hence the latter family is complete by Lemma 2.12. $\qquad\square$

    Finally, we remark that in order to obtain Cor. 7.7 from Thm. 7.5 and Thm. 7.6 one can check that the projections $f_n \leq g_m$ occuring in the proofs of these theorems actually yield relations $N_n f_n(X) = g_m(a)$, where the components of a are either indeterminates, 0 or 1, and the factor N_n is a nonzero integer. Moreover, m, N_n, and a are computable in polynomial time from n. Thus we get reductions between the corresponding problems to evaluate f_n and g_m at 0, 1-values. Moreover, one can obtain from Thm. 7.10 that the problem to compute $\mathrm{im}_{\lambda^{(n)}}(A) + n^{2n}$ from $A \in \{0,1\}^{n \times n}$ is contained in #P, for any polynomial time computable map $n \mapsto \lambda^{(n)}$.

8

Separation Results and Future Directions

We develop techniques for proving that specific families are not p-definable. As an application, we clarify the relation between several complexity classes: VP is strictly contained in VQP, but VQP is not contained in VNP. These separations are proven by means of specific p-families. Finally, we establish a connection between Valiant's model and the Blum-Shub-Smale model, and outline possible directions for deepening our understanding of this connection.

8.1 Specific Families Which Are Not p-Definable

The techniques of algebraic complexity theory sometimes allow to prove non-trivial lower bounds on the complexity of specific polynomials. Consider for instance the univariate polynomial $g_n = \sum_{j<n} \sqrt{p_j}\, T^j$, where p_j is the jth prime number. One can show that $L(g_n)^2 \geq cn/\log n$, where $c > 0$ is a constant. (For an elementary proof see [21, Cor. 9.4].) This example can be readily adapted to fit Valiant's framework as follows. Let $j(e) = \sum_{s=1}^{n} e_s 2^{s-1}$ be the natural number with binary representation $e \in \{0,1\}^n$ and put $X^e = X_1^{e_1} \cdots X_n^{e_n}$. Note that $\{j(e) \mid e \in \{0,1\}^n\} = \{0, 1, \ldots, 2^n - 1\}$. Consider the multivariate polynomial

$$(8.1) \qquad f_n = \sum_{e \in \{0,1\}^n \setminus 0} \sqrt{p_{j(e)}} X^e \ .$$

We claim that the family (f_n) is not p-computable. Indeed, we have

$$g_{2^n}(T) = f_n(T^{2^0}, T^{2^1}, \ldots, T^{2^{n-1}}) \ .$$

Hence $L(g_{2^n}) \leq L(f_n) + n - 1$. The assertion follows from the above mentioned lower bound, which implies $L(g_{2^n})^2 \geq c2^n/n$.

This makes it of course very unlikely that (f_n) is p-definable. The goal of this section is to show that the techniques of algebraic complexity, as explained in [21, Chap. 9], can be extended to prove that (f_n) and related p-families are indeed not p-definable. This result confirms our belief that the separation of VP and VNP is impossible by the present day techniques of algebraic complexity. In the next section, we will apply our techniques to clarify the relations between the complexity classes VP, VQP, and VNP.

In the sequel, p-definability always refers to the algebraic closure $\overline{\mathbb{Q}}$ of $\mathbb{Q}$. We remark that by Cor. 4.2, a p-family with coefficients in $\overline{\mathbb{Q}}$ is p-definable over $\overline{\mathbb{Q}}$ iff it is p-definable over any field extension of $\overline{\mathbb{Q}}$.

The following theorem can be used to show that certain p-families of polynomials with algebraic coefficients of high degree are not p-definable. Its statement and proof are similar to Thm. 9.15 in [21] due to Heintz and Sieveking [49].

Theorem 8.1 *Assume (f_n) is a p-family over $\overline{\mathbb{Q}}$ and let $N(n)$ denote the degree of the field extension generated by the coefficients of f_n over $\mathbb{Q}$. We assume the following:*

(1) *The map $\mathbb{N} \to \mathbb{N}$, $n \mapsto \lceil \log N(n) \rceil$ is not p-bounded.*

(2) *For all n, there is a system G_n of rational polynomials of degree at most $D(n)$ with finite zeroset, containing the coefficient system of f_n, and such that $n \mapsto \lceil \log D(n) \rceil$ is p-bounded.*

Then the family (f_n) is not p-definable.

Let us draw a first conclusion.

Corollary 8.2 *The family (f_n) defined by equation (8.1) is not p-definable.*

Proof. We have $N(n) = [\mathbb{Q}(\sqrt{p_j} \mid 1 \le j < 2^n) : \mathbb{Q}] = 2^{2^n - 1}$ (see for instance [21, Lemma 9.20]), hence $n \mapsto \log N(n)$ is not p-bounded. Moreover, the system of polynomials $G_n = \{Z_j^2 - p_j \mid 1 \le j < 2^n\}$ satisfies the requirements of Thm. 8.1. $\qquad\square$

In order to prove Thm. 8.1, we need the following multivariate version of the representation theorem due to Strassen [105] and Schnorr [95]. It can be shown as Prop. 9.11 in [21], with considerable simplifications due to the fact that we do not allow divisions. (Compare also Lemma 5.30.) We will omit the proof.

$L_{ns}(f)$ stands for the nonscalar complexity to compute f from the variables X_i and constants in $\overline{\mathbb{Q}}$ without divisions. (That is, only nonscalar multiplications are counted.) We recall that the weight $\mathrm{wt}(f)$ and the height $\mathrm{ht}(f)$ of an integer polynomial f are defined as the sum and the maximum of the absolute values of its coefficients, respectively.

Proposition 8.3 *Let $m, r \ge 1$ be given and set $q := r^2 + 2mr + m + 1$. For $\mu \in \mathbb{N}^m$, $|\mu| \le 2^r$, there exist polynomials $F_\mu \in \mathbb{Z}[Z_1, \dots, Z_q]$ such that*

$$\deg F_\mu \le r|\mu| + 1, \quad \sum_{|\mu| \le 2^r} \mathrm{wt}(F_\mu) \le 3(mr)^{2^r} \ ,$$

$$\left\{ f \in \overline{\mathbb{Q}}[X_1, \dots, X_m] \mid L_{ns}(f) \le r \right\} = \left\{ \sum_{|\mu| \le 2^r} F_\mu(\zeta) X^\mu \mid \zeta \in \overline{\mathbb{Q}}^q \right\} \ .$$

Proof. (of Thm. 8.1) Let (f_n) satisfy the assumptions of Thm. 8.1. We assume that (f_n) is p-definable. Then there exists some p-computable family (g_n), say $g_n \in \overline{\mathbb{Q}}[X_1, \ldots, X_{m(n)}]$, such that

$$f_n(X_1, \ldots, X_{v(n)}) = \sum_{e \in \{0,1\}^{m(n)-v(n)}} g_n(X_1, \ldots, X_{v(n)}, e_{v(n)+1}, \ldots, e_{m(n)}) \ .$$

Note that the number of variables $m(n)$, the complexity $r(n) := L_{ns}(g_n)$, and the degree $d(n) := \deg g_n$ are p-bounded functions of n.

Let $F_\mu^{(n)} \in \mathbb{Z}[Z_1, \ldots, Z_{q(n)}]$ be the polynomials of Prop. 8.3 assigned to the values $m = m(n)$, $r = r(n)$, and denote by U_n the Zariski closure of

$$\left\{ \sum_{|\mu| \leq d(n)} F_\mu^{(n)}(\zeta) X^\mu \ \Big| \ \zeta \in \overline{\mathbb{Q}}^{q(n)} \right\}$$

in the vector space V_n of polynomials of degree at most $d(n)$ in $X_1, \ldots, X_{m(n)}$. Note that $g_n \in U_n$. Let π_n be the linear map $V_n \to V_n, g \mapsto f$, where is f is defined by g in the same way as f_n is defined by g_n. Obviously, f_n is contained in the Zariski closure W_n of the image of U_n under π_n. Observe that W_n is the Zariski closure of the image of a polynomial map $\overline{\mathbb{Q}}^{q(n)} \to \overline{\mathbb{Q}}^M$ given by integer polynomials of degree at most $r(n)d(n) + 1$. From this we deduce the following properties of W_n:

(1) $\dim W_n \leq q(n)$,

(2) $\deg W_n \leq (r(n)d(n) + 1)^{q(n)}$,

(3) W_n is defined over $\mathbb{Q}$, i.e., it is stable under the Galois group $\mathrm{Gal}(\overline{\mathbb{Q}}/\mathbb{Q})$.

(To show (2) use [21, Thm. 8.48]; for (3) use [21, Lemma 9.12(3)].)

Let G_n be a system of polynomials of degree at most $D(n)$ as in the statement of Thm. 8.1. Thus $\lceil \log D(n) \rceil$ is supposed to be p-bounded, and the zeroset $Z(G_n)$ is finite and contains the coefficient system α_n of f_n. The set $W_n \cap Z(G_n)$ is stable under the action of $\mathrm{Gal}(\overline{\mathbb{Q}}/\mathbb{Q})$, hence the zeroset $Z(G_n)$ contains the orbit of α_n. By [21, Lemma (9.12)(2)], the cardinality of this orbit equals the degree $N(n)$ of the field generated by the coefficients of f_n over $\mathbb{Q}$. We conclude with [21, Lemma 9.14] that

$$|W_n \cap Z(G_n)| \leq D(n)^{\dim W_n} \deg W_n \ .$$

Using the estimates above this implies that

$$N(n) \leq (D(n)(r(n)d(n) + 1))^{q(n)} \ .$$

Hence $\lceil \log N(n) \rceil$ is a p-bounded function of n, contradicting our assumptions. $\qquad \square$

We give two further applications of Thm. 8.1. For proofs compare Cor. 9.16 and Cor. 9.21 of [21].

Corollary 8.4 *Let $r \in \mathbb{Q} \setminus \mathbb{N}$ and $j(e) := \sum_{s=0}^{n-1} e_s 2^{s-1}$. The following p-families (f_n) are not p-definable $(i = \sqrt{-1})$:*

$$f_n \;=\; \sum_{e \in \{0,1\}^n \setminus 0} \exp\left(\frac{2\pi i}{j(e)}\right) X^e \;,$$

$$f_n \;=\; \sum_{e \in \{0,1\}^n} \exp\left(\frac{2\pi i}{2^{j(e)}}\right) X^e \;,$$

$$f_n \;=\; \sum_{e \in \{0,1\}^n} j(e)^r X^e \;.$$

We proceed with a result that can be used to show that specific families with integer coefficients are not p-definable. The proof techniques are borrowed from Strassen [105].

Proposition 8.5 *Let (f_n) be a p-family over $\mathbb{Q}$ and $t \colon \mathbb{N} \to \mathbb{N}$ be a function satisfying $n^{\log n} \le t(n) \le 2^n$, such that for all n each of the integers 2^{2^j}, for $0 \le j < t(n)$, appears as a coefficient of f_n. Then (f_n) is not p-definable.*

As a corollary we obtain for instance the following (take $t(n) = 2^n$).

Corollary 8.6 *The family (f_n) defined by*

$$f_n = \sum_{e \in \{0,1\}^n} 2^{2^{j(e)}} X^e$$

is not p-definable.

Proof. (of Prop. 8.5) We suppose that (f_n) is p-definable. Hence there is a p-computable family (g_n) such that $f_n(X) = \sum_e g_n(X, e)$. Arguing as in the proof of Thm. 8.1 (and using the same notation), we see that there exist polynomials $F_\mu^{(n)}$ in $\mathbb{Z}[Z_1, \ldots, Z_{q(n)}]$ which satisfy for $|\mu| \le \deg g_n = n^{O(1)}$

$$\deg F_\mu^{(n)} = n^{O(1)}, \;\; \log \mathrm{wt}(F_\mu^{(n)}) = 2^{n^{O(1)}} \;,$$

and such that for some $\zeta \in \overline{\mathbb{Q}}^{q(n)}$

$$g_n = \sum_\mu F_\mu^{(n)}(\zeta)\, X_1^{\mu_1} \cdots X_{m(n)}^{\mu_{m(n)}} \;.$$

This implies that

$$f_n = \sum_{\nu \in \mathbb{N}^{v(n)}} Q_\nu^{(n)}(\zeta)\, X_1^{\nu_1} \cdots X_{v(n)}^{\nu_{v(n)}} \;,$$

where the polynomials $Q_\nu^{(n)}$ are obtained as a sum of at most $2^{m(n)}$ polynomials $F_\mu^{(n)}$. Hence we have estimates $\deg Q_\nu^{(n)} \le \alpha(n)$ and $\log \mathrm{wt}(Q_\nu^{(n)}) \le 2^{\beta(n)}$, where $\alpha(n)$ and $\beta(n)$ are p-bounded functions of n.

Put $s(n) := \lfloor t(n)/n \rfloor$. By our assumption, we have for each n an index set $N_n \subseteq \mathbb{N}^{\nu(n)}$ and a bijection $\{0, 1, \dots, s(n) - 1\} \to N_n$, $\sigma \mapsto \nu(\sigma)$ such that

$$Q_{\nu(\sigma)}^{(n)}(\zeta) = 2^{2^{\sigma n + 1}}$$

for all $\sigma < s(n)$.

Lemma 9.28 of [21] implies the existence of forms $H_n \in \mathbb{Z}[Y_\nu \mid \nu \in N_n] \setminus 0$ with $\mathrm{ht}(H_n) \le 3$, $\deg H_n \le D(n)$, and such that

$$H_n(Q_\nu^{(n)} \mid \nu \in N_n) = 0 \ ,$$

provided

$$D(n)^{s(n) - q(n) - 2} > \alpha(n)^{q(n)} s(n)^{s(n)} 2^{\beta(n)} \ .$$

We claim that this inequality is satisfied for $D(n) = 2^n - 1$ if n is sufficiently large. For this it suffices to show that $s(n) \log D(n)/s(n)$ grows faster than any polynomial function. But $D(n)/s(n) \ge n(2^n - 1)/2^n \ge 2$, as $t(n) \le 2^n$. On the other hand, $s(n)$ is growing faster than any polynomial, since we assume that $t(n) \ge n^{\log n}$.

We may write

$$H_n = \sum_\epsilon \lambda_\epsilon \prod_{\nu \in N_n} Y_\nu^{\epsilon_\nu} \ ,$$

where λ_ϵ are integers of absolute value at most three. Then we have

$$
\begin{aligned}
0 \ = \ H_n(Q_\nu^{(n)} \mid \nu \in N_n) \ &= \ \sum_\epsilon \lambda_\epsilon \prod_\sigma 2^{\epsilon_{\nu(\sigma)} 2^{\sigma n + 1}} \\
&= \ \sum_\epsilon \lambda_\epsilon \, 4^{\sum_\sigma \epsilon_{\nu(\sigma)} (2^n)^\sigma} \ .
\end{aligned}
$$

Note that the exponents $\sum \epsilon_{\nu(\sigma)} (2^n)^\sigma$ are pairwise distinct (2^n-adic representation, $\epsilon_\nu < 2^n$). Hence the above is the 4-adic representation of an integer (recall $|\lambda_\epsilon| \le 3$). Therefore, we conclude $\lambda_\epsilon = 0$ for all ϵ, which contradicts $H_n \ne 0$. $\qquad\square$

Remark 8.7 Let (f_n) satisfy the assumptions of Thm. 8.1. Then it is true that (f_n) is not p-definable relative to any p-family h of rational polynomials. The proof is similar to the one given here: one works with a generalization of Prop. 8.3, which can be essentially found in Lemma 5.30.

8.2 Separations for the Complexity Class VQP

In Sect. 2.5, we have introduced the complexity class VQP of qp-computable families in connection with a completeness result for the determinant family.

Von zur Gathen [41] asked for a proof that VP is strictly contained in VQP. Our first goal is to provide such a proof. We work over a field k of characteristic zero.

We define the class VQP^i for $i \in \mathbb{N}$ as the set of all p-families (f_n) satisfying $L(f_n) \le n^{O(\log^i n)}$. We have $\mathrm{VP} = \mathrm{VQP}^0$ and inclusions $\mathrm{VQP}^i \subseteq \mathrm{VQP}^{i+1}$. The union of the classes VQP^i equals VQP.

For $i \ge 1$ we introduce the p-family $f^i = (f^i_n)$ defined by

$$f^i_n = \sum_\mu 2^{2^{j_n(\mu)}} X_1^{\mu_1} \cdots X_{m(n)}^{\mu_{m(n)}} \ ,$$

where $j_n(\mu) := \sum_{j=1}^{m(n)} \mu_j n^{j-1}$ and $m(n) := m_i(n) := \lceil \log^i n \rceil$. The sum is over all systems $\mu \in \{0, 1, \ldots, n-1\}^{m(n)}$.

An easy induction on m shows that any polynomial $f \in k[X_1, \ldots, X_m]$ satisfying $\deg_{X_j} f < n$ for all j has complexity $L(f) \le 2n^m - 2$. This observation implies that the family f^i is contained in VQP^i.

Corollary 8.8 *The family f^i is contained in VQP^i but not in VQP^{i-1}. In particular, the inclusions $\mathrm{VQP}^i \subseteq \mathrm{VQP}^{i+1}$ are strict, and we have the separation $\mathrm{VP} \neq \mathrm{VQP}$.*

Proof. For fixed $i > 0$ let $m(n) = m_i(n)$ and consider the univariate polynomial $g_n := \sum_j 2^{2^j} T^j$, where the sum is over all $0 \le j < n^{m(n)}$. This polynomial is related to f_n by the equation

$$g_n = f_n(T^{n^0}, T^{n^1}, \ldots, T^{n^{m(n)-1}}) \ .$$

Since $L(T^n) \le 2\log n$, we have $L(g_n) \le L(f_n) + 2m(n)\log n$. On the other hand, it is known that (cf. [21, Cor. 9.32])

$$L^2(g_n) \ge c \frac{n^{m(n)}}{m(n)\log n} \ ,$$

where $c > 0$ is a constant. This implies $(f_n) \notin \mathrm{VQP}^{i-1}$. $\qquad\square$

In Sect. 2.5 we mentioned the Valiant's extended hypothesis, which claims that VNP is not contained in VQP. As a corollary of Prop. 8.5 we can now prove that the reverse inclusion does not hold.

Corollary 8.9 *The family f^1 is qp-computable, but not p-definable. Hence VQP is not contained in VNP.*

Proof. We know already that $f^1 \in \mathrm{VQP}^1$. Prop. 8.5 immediately implies that f^1 is not p-definable (take $t(n) = n^{\lceil \log n \rceil}$). $\qquad\square$

Up to now, DET is the only known VQP-complete family. We have the following conjecture, which is inspired by the reductions in Cook [26] and von zur Gathen [40].

Conjecture 8.1 The following p-families (f_n) are VQP-complete (X and X_i stand for n by n matrices with independent indeterminate entries and Tr denotes the trace):

$$\begin{aligned}
f_n &= \mathrm{Tr}(X^n) & \text{(matrix power)} \\
f_n &= \mathrm{Tr}(X_1 \cdots X_n) & \text{(iterated matrix product)} \\
f_n &= \mathrm{Tr}(\det(X)X^{-1}) & \text{(adjoint matrix)}
\end{aligned}$$

8.3 Possible Connections to Univariate Polynomials

Shub and Smale [98] discovered an astonishing connection between the complexity of univariate integer polynomials and the P-NP-hypothesis in the BSS-model. We will indicate that there might be a similar connection between the complexity of univariate polynomials and Valiant's hypothesis.

In order to explain the relation to the BSS-model, we introduce a modified complexity notion. For a univariate polynomial $f \in \mathbb{Z}[T]$, let the *tau-complexity* $\tau(f)$ be the minimal number of arithmetic operations $+, -, *$ sufficient to compute f from T and the constant 1. Note that if we allowed for arbitrary constants in the underlying field, then we would get our usual complexity measure $L(f)$. It is clear that $L(f) \leq \tau(f)$.

The τ-*conjecture* claims the following connection between the number $z(f)$ of distinct integer roots of $f \in \mathbb{Z}[T]$ and the complexity $\tau(f)$:

$$(8.2) \qquad\qquad z(f) \leq (1 + \tau(f))^c \ ,$$

where $c > 0$ is a universal constant. Shub and Smale [98] proved that the τ-conjecture implies P $\neq$ NP over the field of complex numbers. In fact, their proof shows that in order to draw this conclusion, it suffices to prove the existence of some $\epsilon > 0$ such that the lower bound $\tau(f_n) \geq n^\epsilon$ holds for all nonzero multiples f_n of the so-called *Pochhammer polynomials* $P_n = \prod_{j=1}^{n}(T - j)$, for all n.

For discussing this result, it is convenient to introduce the following terminology: We call a sequence (g_n) of rational univariate polynomials satisfying $\deg g_n = n$ easy to compute iff there is some $c > 0$ such that $L(g_n) \leq \log^c n$ for infinitely many n. Otherwise, we call (g_n) *hard to compute*.

We do not know whether the sequence of Pochhammer polynomials (P_n) is hard to compute. However, Heintz and Morgenstern [48] made the following interesting observation: consider the factorization

$$P_n(T^2) = \underbrace{\prod_{j=1}^{n}(T - \sqrt{j})}_{Q_n} \cdot \underbrace{\prod_{j=1}^{n}(T + \sqrt{j})}_{\overline{Q}_n} \ .$$

Then each of the sequences (Q_n) and $(\overline{Q}_n)$ is hard to compute.

It is an outstanding open problem to produce "specific" sequences of polynomials with coefficients of p-bounded bitsize in n, that are hard to compute. We think that Valiant's theory can give some explanation of the intricacy of this problem. Let us formalize the notion of a specific family of univariate polynomials by means of the concept of p-definability as follows.

Let $g_n = \sum_{j=0}^n \alpha_{n,j} T^j \in \mathbb{Q}[T]$ be of degree n for all $n \in \mathbb{N}$. We assign to g_{2^n-1} the multivariate polynomial

$$f_n := \sum_{e \in \{0,1\}^n} \alpha_{2^n-1, j(e)} X_1^{e_1} \cdots X_n^{e_n} \ ,$$

where $j(e) := \sum_{s=1}^n e_s 2^{s-1}$. Note that $g_{2^n-1}(T) = f_n(T^{2^0}, \ldots, T^{2^{n-1}})$, which implies $L(g_{2^n-1}) \le L(f_n) + n - 1$.

Definition 8.10 We call (g_n) *easily definable* iff (f_n) is p-definable.

For instance, Cor. 8.6 implies that the family $(\sum_{j=0}^n 2^{2^j} T^j)$ is not easily definable. The following remark provides us with plenty of examples of easily definable families. It is an immediate consequence of Valiant's criterion 2.20.

Lemma 8.11 *Assume that the coefficient map* $\mathbb{N}^2 \to \mathbb{N}, (n, j) \mapsto \alpha_{n,j}$ *is contained in the complexity class* #P/poly *(the natural numbers are to be encoded in binary). Then* (g_n) *is easily definable.*

Our interest in easily definable families is motivated by the following observation.

Remark 8.12 If there exists an easily definable, but hard to compute sequence of univariate polynomials, then Valiant's hypothesis is true over the rationals.

We raise the following questions.

Problem 8.1   (1) Is the sequence of Pochhammer polynomials easily definable?

(2) Do the Taylor approximations $\sum_{j=1}^n \frac{1}{j} T^j$ and $\sum_{j=1}^n \frac{1}{j!} T^j$ of the logarithm and of the exponential function, respectively, form easily definable families?

A positive answer to the first question would show that Valiant's hypothesis over the rationals is a consequence of the τ-conjecture (with τ replaced by L). An affirmative answer to the second question would explain our difficulty to prove that the Taylor approximations of log and exp are hard to compute.

8.4   Connections to the BSS-Model

In this monograph, we have undertaken a thorough investigation of Valiant's algebraic theory of NP-completeness. An important source of our inspiration was the now classical theory of NP-completeness, originating in the work of Cook [25] and Karp [60], as well as the recent, more general developments of a theory of complexity and computation over the reals (and other algebraic structures) initiated by the work of Blum, Shub, and Smale [12] (BSS-model).

In each of these theories, we have a fundamental conjecture claiming that the corresponding classes P and NP are different. We have called them Cook's hypothesis, BSS hypothesis, and Valiant's hypothesis, respectively. The latter two depend on the underlying field. Over the field $\mathbb{F}_2$ with two elements, Cook's hypothesis and the BSS-hypothesis coincide.

The proof of each of these conjectures seems to be a very hard mathematical problem. However, we can say something about the interrelations between the different P-NP-hypotheses. Our knowledge about the connections of Cook's hypothesis with the BSS hypothesis and Valiant's hypothesis over the complex numbers is depicted in Fig. 1.1 of the introduction. It has been shown independently by several people, including ourselves, that the nonuniform version of Cook's hypothesis $P/poly \neq NP/poly$ implies $P \neq NP$ over $\mathbb{C}$. (For a proof see [31].) Whether the corresponding implication is true over the reals (where we have $\leq$-branchings) is a major open problem in the BSS-theory. In Chap. 4, we have proved that the nonuniform version of Cook's hypothesis implies Valiant's hypothesis over $\mathbb{C}$, under the generalized Riemann hypothesis. Observe that both of these known implications go from the bit-model to an algebraic model. This is in accordance with our general belief that lower bounds in the algebraic model are easier to obtain.

We have the following conjecture, which, loosely speaking, states that if the permanent is intractable, then solving systems of polynomial equations is intractable as well.

Conjecture 8.2 Valiant's hypothesis implies the BSS hypothesis over the complex numbers.

In what follows, we will prove some weaker implications and clarify the problems that have to be overcome in order to settle this conjecture.

The complexity classes P and NP in the BSS-model consist of decision problems, whereas Valiant's classes VP and VNP formalize computational problems. In order to compare these two models, one has to investigate the relationship between the complexities of algebraic decision and computational problems.

Consider a nonzero polynomial g in n variables over the field $k = \mathbb{R}$ or $k = \mathbb{C}$. Let $C(g)$ denote the complexity to test membership of given points in k^n to the hypersurface $g = 0$ by means of algebraic computation trees. We count all arithmetic operations and tests. Moreover, the trees may branch according to $\leq$-tests in the case $k = \mathbb{R}$. (For formal definitions we refer to [21].) It is clear that $C(g) \leq L(g) + 1$.

Without loss of generality we may assume that the computation trees are division-free. Namely, we can eliminate divisions by representing rational functions a/b by pairs (a, b) of polynomials. An arithmetic operation on two rational functions a/b and a_1/b_1 can be clearly simulated by 4 additions, subtractions, or multiplications of the polynomials a, b, a_1, b_1. Moreover, we have $a/b \geq 0$ iff $ab \geq 0$. Therefore, the restriction of no divisions increases the complexity at most by a factor of 4.

The following lemma is folklore. (See for instance [22], or the canonical path argument in [11].)

Lemma 8.13 *There exists a nonzero multiple f of g such that $L(f) \leq C(g)$. Over $\mathbb{C}$ this is true without additional assumption, over $\mathbb{R}$ we have to require that g is irreducible.*

To see that the irreducibility is necessary over $\mathbb{R}$, consider a univariate polynomial g with d roots that are algebraically independent over $\mathbb{Q}$. Then $L(f) \geq d$ for all nonzero multiples f of g, but we can decide for a given $x \in \mathbb{R}$ whether $g(x) = 0$ with $O(\log d)$ tests using a bisection method (compare [22]).

If we could assume in Lemma 8.13 that $L(g)$ is polynomially bounded by $L(f)$, then a proof of our Conjecture 8.2 would be supplied by the following argument similar as in Heintz and Morgenstern [48].

Consider the following decision problem PERMUT: Given a matrix $x \in \mathbb{C}^{n \times n}$ and $z \in \mathbb{C}$, is there a permutation $\pi \in S_n$ such that $zx_{1,\pi(1)} \cdots x_{n,\pi(n)}$ equals 1? This problem is contained in the BSS complexity class NP over $\mathbb{C}$. To directly see this, we may express the above condition by

$$\exists y \in \{0,1\}^{n \times n} : z \prod_{i=1}^{n} \prod_{j=1}^{n} x_{i,j} y_{i,j} = 1 \ \wedge \ h_n(y) = 1 \ ,$$

where h_n is the product of the polynomials α_n and β_n encountered in the proof of Lemma 2.6. Note that $L(h_n)$ is p-bounded in n. (This also shows that the above problem is contained in the class DNP of digital nondeterminism; for a definition see [11].) We introduce the polynomial

$$g_n := \prod_{\pi \in S_n} (1 - ZX_{1,\pi(1)} \cdots X_{n,\pi(n)}) \ .$$

Then PERMUT is nothing but the problem to check membership to the hypersurface $g_n = 0$.

Assume now that P = NP over $\mathbb{C}$ (the assumption P = DNP would suffice). Then PERMUT would be contained in P, hence $C(g_n)$ would be a p-bounded function of n. By Lemma 8.13 there is multiple f_n of g_n for each n such that $L(f_n) \leq C(g_n)$. If $L(g_n)$ were polynomially bounded by $L(f_n)$, then we would have that $L(g_n)$ is a p-bounded function of n. The expansion of g_n according to powers of Z starts as follows: $g_n = 1 - Z\,\mathrm{PER}_n(X) + O(Z^2)$. A variant of Lemma 2.14 on homogeneous parts implies $L(\mathrm{PER}_n) \leq 4L(g_n)$, hence $L(\mathrm{PER}_n)$ would be a p-bounded function of n. The completeness of PER implies that $\mathrm{VP} = \mathrm{VNP}$.

This proof attempt clearly shows that the relation of the complexity of a polynomial to those of its factors is crucial for our question. Unfortunately, there exist polynomials f having factors with a complexity exponential in the complexity of f. This was first discovered by Lipton and Stockmeyer [75]. For instance, consider $f_n = X^{2^n} - 1 = \prod_{j<2^n}(X - \zeta^j)$, where $\zeta = \exp(2\pi i/2^n)$. By repeated squaring we get $L(f_n) \leq n+1$. On the other hand, one can prove that for almost all $M \subseteq \{0, 1, \ldots, 2^n - 1\}$ we have

$$L\Big(\prod_{j \in M} (X - \zeta^j)\Big) \geq \frac{2^{n/2}}{\sqrt{3n}}$$

(cf. [21, Ex. 9.8]).

We conjecture that this cannot happen for factors of moderate degree.

Conjecture 8.3 Let g be a factor of a polynomial f in n variables over a field k of characteristic zero. Then we have with a universal constant $c > 0$ that

$$L(g) \leq (L(f) + \deg g + n)^c \ .$$

This means that the complexity of the factor g is polynomially bounded in the complexity of f, the degree of g, and the number of variables n.

Recall Thm. 2.21 from Chap. 2. There it was shown that the complexity of the factor g is polynomially bounded in the complexity of f, the degree of f, and the number of variables n. Conjecture 8.3 is considerably supported by the following extension of Thm. 2.21. (We remark that a similar result had been independently obtained by Kaltofen [56].)

Theorem 8.14 *Assume $f = g^e h$ with polynomials $g, h \in k[X_1, \ldots, X_n]$ which are relatively prime. Let $d \geq 1$ be the degree of g. We suppose that k is a field of characteristic zero. Then we have*

$$L(g) = O(M(d)^2 M(de)[L(f) + n + d \log e]) \ .$$

In particular, the complexity of the factor g is polynomially bounded in the complexity of f, the degree d, the multiplicity e, and the number of variables n.

Therefore, the only way Conjecture 8.3 could fail is when a huge multiplicity e occurs. We are working towards a result stating that for approximative complexity, the dependence on the multiplicity e in the above upper bound can be removed. (See [16], where also a proof of Thm. 8.14 can be found.)

The complexity class PAR is defined as the class of decision problems that can be decided by parallel BSS-machines that work in polynomial time using an exponential number of processors. This class was introduced in [29] and can be interpreted as an analogue of the classical complexity class PSPACE. (For formal definitions see [11].)

Remark 8.15 If Conjecture 8.3 were true, then the following conclusions could be drawn.

(1) Let g be an irreducible polynomial in n variables of degree d over the reals or complex numbers. Then the complexity $L(g)$ is polynomially bounded in the test complexity $C(g)$, the degree d, and n.

(2) Valiant's hypothesis implies $P \neq PAR$ over the reals.

The first claim is an immediate consequence of Lemma 8.13.

The proof of the second claim runs as follows: Consider the following decision problem PERTEST over the reals: given a matrix $x \in \mathbb{R}^{n \times n}$, test whether $\mathrm{per}(x) = 0$. It is clear that this problem can be solved by parallel machines in polynomial time using an exponential number of processors. Hence this decision problem lies in the class PAR. Assume now that $P = PAR$. Then $C(\mathrm{PER}_n)$ is p-bounded. The first part of the above remark implies that $L(\mathrm{PER}_n)$ is p-bounded (note that PER_n is irreducible). Hence $VNP = VP$.

We can circumvent Conjecture 8.3 by restricting attention to the the so-called *weak BSS-model* introduced by Koiran [66]. In this model, a different cost function is considered, which takes into account the degrees as well as the bitsizes of the coefficients of all intermediate results. The corresponding weak complexity classes are denoted by P_w, NP_w, and PAR_w. It is almost immediate from the definition that P_w is strictly contained in P and that $NP_w = NP$. (For details see [11].)

Corollary 8.16 *Valiant's hypothesis implies* $P_w \neq PAR_w$.

Proof. The problem PERTEST lies in the class PAR_w. Indeed, the obvious parallel machine solving this problem only computes polynomials of p-bounded degree and with integer coefficients, which are of p-bounded bitsize in the input.

Assume now that $P_w = PAR_w$. Then, as in the proof of Lemma 8.13, we see that for each n there is a nonzero multiple f_n of the permanent polynomial PER_n with p-bounded complexity. Moreover, because of the weak cost measure, f_n has *p-bounded degree*. Thm. 2.21 implies that $L(\mathrm{PER}_n)$ is p-bounded as well, which contradicts Valiant's hypothesis. $\qquad\square$

In Koiran [66] and Cucker et al. [33] the Boolean parts of weak BSS-classes have been determined: they showed that

$$BP(P_w) = P/\mathrm{poly}, \quad BP(PAR_w) = PSPACE/\mathrm{poly} \ .$$

This gives the implication

$$P/\mathrm{poly} \neq PSPACE/\mathrm{poly} \implies P_w \neq PAR_w \ .$$

We can strengthen this by relating "$P_w = PAR_w$" to a collapse of nonuniform discrete *parallel* complexity classes.

Corollary 8.17 $\mathrm{NC}^3/\mathrm{poly} \neq \mathrm{PSPACE}/\mathrm{poly} \implies \mathrm{P}_w \neq \mathrm{PAR}_w$.

Proof. (Sketch) Assume $\mathrm{P}_w = \mathrm{PAR}_w$. Then there is a BSS-machine M which solves the problem PERTEST in weak polynomial time. Let $a_1, \ldots, a_m \in \mathbb{R}$ be the machine constants. As in the proof of Cor. 8.16, we see that for each n there is a nonzero multiple f_n of the permanent polynomial PER_n satisfying the following: there is a straight-line program of p-bounded size computing f_n from the variables $X_{i,j}$ and the constants a_j such that all intermediate results are integer polynomials in $X_{i,j}$ and a_j of p-bounded degree, and with coefficients of p-bounded bitsize. By arguing as in Prop. 4.1(iii), we may assume that $a_1, \ldots, a_m$ are algebraically independent. (This way, we increase the complexity at most by a constant factor $O(m^3)$.) Then we substitute $a_j \mapsto 0$ and thus eliminate the constants (note that we do not have divisons).

Thm. 2.21 implies that the complexity of the factor PER_n is p-bounded in n. By checking the proof of this theorem, we see that for each n, there is a straight-line program of p-bounded size computing PER_n and having the property that all intermediate results are integer polynomials in $X_{i,j}$ of p-bounded degree, and with coefficients of p-bounded bitsize.

By Thm. 2.23, we can transform this circuit into one, which additionally has depth $O(\log^2 n)$. (One has to check that the coefficient growth of the intermediate results remains p-bounded under this transformation.)

We can simulate this algebraic circuit on $0,1$-inputs by a Boolean circuit of polynomial size and depth $O(\log^3 n)$. (Recall that addition and multiplication of ℓ-bit integers can be done by Boolean circuits of size $\ell^{O(1)}$ and depth $O(\log \ell)$.) This implies that the problem to compute the permanent of $0,1$-matrices lies in the complexity class $\mathrm{FNC}^3/\mathrm{poly}$. In particular, we get the inclusion $\mathrm{P} \subseteq \mathrm{NC}^3/\mathrm{poly}$.

On the other hand, by the above mentioned result of [33] we get from our assumption $\mathrm{P}_w = \mathrm{PAR}_w$ that $\mathrm{P}/\mathrm{poly} = \mathrm{PSPACE}/\mathrm{poly}$. Altogether, we obtain $\mathrm{NC}^3/\mathrm{poly} = \mathrm{PSPACE}/\mathrm{poly}$. $\qquad\square$

References

[1] L. Adleman. Two theorems on random polynomial time. In *Proc. 19th IEEE Symp. on Foundations of Comp. Science*, pages 75–83, 1978.

[2] E. Allgower and K. Georg. Numerical path following. In P. Ciarlet and J.-L. Lions, editors, *Handbook of Numerical Analysis*, volume V, pages 3–207. North-Holland, 1997.

[3] T. Baker, J. Gill, and R. Solovay. Relativizations of the $P =?NP$ question. *SIAM J. Comp.*, 4:431–442, 1975.

[4] J. L. Balcázar, J. Díaz, and J. Gabarró. *Structural Complexity I*. Springer Verlag, 1988.

[5] A.I. Barvinok. Computational complexity of immanants and representations of the full linear group. *Functional Analysis Applications*, 24:144–145, 1990.

[6] A.I. Barvinok. Two algorithmic results for the traveling salesman problem. *Mathematics of Operations Research*, 21:65–84, 1996.

[7] S. Ben-David, K. Meer, and C. Michaux. A note on non-complete problems in NP_R. Preprint, 1996.

[8] C.H. Bennett and J. Gill. Relative to a random oracle A, $P^A \neq NP^A \neq \mathrm{co} - NP^A$ with probability 1. *SIAM J. Comp.*, 10:96–113, 1981.

[9] L.C. Biedenharn and J.D. Louck. *Angular Momentum in Quantum Physics: Theory and Application; The Racah-Wigner Algebra in Quantum Theory*. Number 8 and 9 in Encyclopedia of Mathematics and its Applications. Addison-Wesley, 1981.

[10] L. Blum, F. Cucker, M. Shub, and S. Smale. Algebraic Settings for the Problem "$P \neq NP$?". In *The mathematics of numerical analysis*, number 32 in Lectures in Applied Mathematics, pages 125–144. Amer. Math. Soc., 1996.

[11] L. Blum, F. Cucker, M. Shub, and S. Smale. *Complexity and Real Computation*. Springer, 1998.

[12] L. Blum, M. Shub, and S. Smale. On a theory of computation and complexity over the real numbers. *Bull. Amer. Math. Soc.*, 21:1–46, 1989.

[13] H. Boerner. *Representations of groups*. Elsevier North-Holland, Amsterdam, 1970.

[14] B. Bollobás. *Graph Theory*. GTM. Springer Verlag, 1979.

[15] R.P. Brent. The complexity of multiprecision arithmetic. In *Proc. Seminar on Compl. of Comp. Problem Solving, Brisbane*, pages 126–165, 1975.

[16] P. Bürgisser. The complexity of factors of multivariate polynomials. In preparation.

[17] P. Bürgisser. The computational complexity to evaluate representations of general linear groups. *SIAM J. Comp.* To appear.

[18] P. Bürgisser. The computational complexity of immanants. *SIAM J. Comp.* To appear.

[19] P. Bürgisser. Cook's versus Valiant's hypothesis. *Theoret. Comp. Sci.*, 235, 1999. In press.

[20] P. Bürgisser. On the structure of Valiant's complexity classes. *Discr. Math. Theoret. Comp. Sci.*, 3:73–94, 1999.

[21] P. Bürgisser, M. Clausen, and M.A. Shokrollahi. *Algebraic Complexity Theory*, volume 315 of *Grundlehren der mathematischen Wissenschaften*. Springer Verlag, 1997.

[22] P. Bürgisser, T. Lickteig, and M. Shub. Test complexity of generic polynomials. *J. Compl.*, 8:203–215, 1992.

[23] O. Chapuis and P. Koiran. Saturation and Stability in the Theory of Computation over the Reals. *Annals of Pure and Applied Logic*. To appear.

[24] M. Clausen. Fast generalized Fourier transforms. *Theoret. Comp. Sci.*, 67:55–63, 1989.

[25] S.A. Cook. The complexity of theorem proving procedures. In *Proc. 3rd ACM STOC*, pages 151–158, 1971.

[26] S.A. Cook. A taxonomy of problems with fast parallel algorithms. *Inf. Control*, 64:2–22, 1985.

[27] B. Courcelle, J.A. Makowsky, and U. Rotics. On the fixed parameter complexity of graph enumeration problems definable in monadic second order logic. *Discrete Applied Math.* To appear.

[28] F. Cucker. $P_R \neq NC_R$. *J. Compl.*, 8:230–238, 1992.

[29] F. Cucker. On the Complexity of Quantifier Elimination: the Structural Approach. *The Computer Journal*, 36:399–408, 1993.

[30] F. Cucker and D.Yu. Grigoriev. On the power of real Turing machines over binary inputs. *SIAM J. Comp.*, 26:243–254, 1997.

[31] F. Cucker, M. Karpinski, P. Koiran, T. Lickteig, and K. Werther. On real Turing machines that toss coins. In *Proc. 27th ACM STOC, Las Vegas*, pages 335–342, 1995.

[32] F. Cucker and P. Koiran. Computing over the reals with addition and order: Higher complexity classes. *J. Compl.*, 11:358–376, 1995.

[33] F. Cucker, M. Shub, and S. Smale. Separation of complexity classes in Koiran's weak model. *Theoret. Comp. Sci.*, 133:3–14, 1994.

[34] T. Emerson. Relativizations of the P=?NP question over the reals (and other ordered rings). *Theoret. Comp. Sci.*, 133:15–22, 1994.

[35] W. Feller. *An introduction to probability theory and its applications*, volume 2. John Wiley & Sons, 1971.

[36] M.E. Fisher. Statistical mechanics of dimers on a plane lattice. *Phys. Rev.*, 124:1664–1672, 1961.

[37] W. Fulton and J. Harris. *Representation Theory*. Number 129 in GTM. Springer Verlag, 1991.

[38] M.R. Garey and D.S. Johnson. *Computers and Intractability: A Guide to the theory of NP-completeness*. W.H. Freeman and Company, New York, 1979.

[39] M.R. Garey, D.S. Johnson, and R.E. Tarjan. The planar Hamilton circuit problem is NP-complete. *SIAM J. Comp.*, 5:704–714, 1976.

[40] J. von zur Gathen. Parallel arithmetic computations: a survey. In *Proc. 12th Symp. Math. Found. Comput. Sci., Bratislava*, number 233 in LNCS, pages 93–112, 1986.

[41] J. von zur Gathen. Feasible arithmetic computations: Valiant's hypothesis. *J. Symb. Comp.*, 4:137–172, 1987.

[42] J. von zur Gathen. Permanent and determinant. *Lin. Alg. Appl.*, 96:87–100, 1987.

[43] J. von zur Gathen and E. Kaltofen. Factoring sparse multivariate polynomials. *J. Comp. Syst. Sci.*, 31:265–287, 1985.

[44] I.M. Gelfand and M.L. Tsetlin. Finite dimensional representations of the group of unimodular matrices. *Dokl. Akad. Nauk SSSR*, 71:825–828, 1950. (in Russian).

[45] M. Giusti and J. Heintz. La détermination des points isolés et de la dimension d'une varieté algébrique peut se faire en temps polynomial. In D. Eisenbud and L. Robbiano, editors, *Proc. Cortona Conference on Computational Algebraic Geometry and Commutative Algebra*. Cambridge University Press, 1993.

[46] J. Grabmeier and A. Kerber. The Evaluation of Irreducible Polynomial Representations of the General Linear Groups and of the Unitary Groups over Fields of Characteristic 0. *Acta Applicandae Mathematicae*, 8:271–291, 1987.

[47] W. Hartmann. On the complexity of immanants. *Linear and Multilinear Algebra*, 18:127–140, 1985.

[48] J. Heintz and J. Morgenstern. On the intrinsic complexity of elimination theory. *Journal of Complexity*, 9:471–498, 1993.

[49] J. Heintz and M. Sieveking. Lower bounds for polynomials with algebraic coefficients. *Theoret. Comp. Sci.*, 11:321–330, 1980.

[50] C.T. Hepler. On the complexity of computing characters of finite groups. Technical Report 94/545/14, Dept. of Computer Science, University of Calgary, 1994.

[51] G.D. James and A. Kerber. *The Representation Theory of the Symmetric Group*. Addison-Wesley, 1981.

[52] M. Jerrum. *On the Complexity of Evaluating Multivariate Polynomials*. Aiken Comput. Lab., Dept. of Computer Science, University of Edinburgh, 1981.

[53] M. Jerrum. Two-dimensional monomer-dimer systems are computationally intractable. *J. Stat. Phys.*, 48:121–134, 1987. Erratum in: J. Stat. Phys. 59:1087–1088, 1990.

[54] M.R. Jerrum and A. Sinclair. Approximating the permanent. *SIAM J. Comp.*, 18:1149–1178, 1989.

[55] D.S. Johnson. A catalog of complexity classes. In J. van Leeuwen, editor, *Handbook of Theoretical Computer Science*, volume A, chapter 2, pages 61–161. Elsevier Science Publishers B. V., 1990.

[56] E. Kaltofen. Single-factor Hensel lifting and its application to the straight-line complexity of certain polynomials. In *Proc. 19th ACM STOC*, pages 443–452, 1986.

[57] E. Kaltofen. Uniform closure properties of *p*-computable functions. In *Proc. 18th ACM STOC*, pages 330–337, 1987.

[58] E. Kaltofen. Factorization of polynomials given by straight-line programs. In S. Micali, editor, *Randomness and Computation*, pages 375–412. JAI Press, Greenwich CT, 1989.

[59] E. Kaltofen and B.M. Trager. Computing with polynomials given by black boxes for their evaluations: greatest common divisors, factorization, separation of numerators and denominators. *J. Symb. Comp.*, 9:301–320, 1990.

[60] R.M. Karp. Reducibility among combinatorial problems. In R.E. Miller and J.W. Thatcher, editors, *Complexity of Computer Computations*, pages 85–104. New York, 1972.

[61] R.M. Karp and R.J. Lipton. Turing machines that take advice. In *Logic and Algorithmic: An international Symposium held in honor of Ernst Specker*, pages 255–273. Monogr. No. 30 de l'Enseign. Math., 1982.

[62] R.M. Karp and V. Ramachandran. Parallel algorithms for shared-memory machines. In J. van Leeuwen, editor, *Handbook of Theoretical Computer Science*, volume A, chapter 17, pages 869–941. Elsevier Science Publishers B. V., 1990.

[63] P.W. Kasteleyn. The statistics of dimers on a lattice I. The number of dimer arrangements on a quadratic lattice. *Physica*, 27:1209–1225, 1961.

[64] P.W. Kasteleyn. Graph Theory and Crystal Physics. In F. Harary, editor, *Graph Theory and Theoretical Physics*, pages 43–110. Academic Press, 1967.

[65] D.G. Kirkpatrick and P. Hell. On the complexity of general graph factor problems. *SIAM J. Comp.*, 12:601–609, 1983.

[66] P. Koiran. A weak version of the Blum-Shub-Smale model. In *Proc. 34th FOCS*, pages 486–495, 1993.

[67] P. Koiran. Hilbert's Nullstellensatz is in the polynomial hierarchy. *J. Compl.*, 12:273–286, 1996.

[68] T. Krick and L. M. Pardo. Une approche informatique pour l'approximation diophantienne. *C. R. Acad. Sc. Paris*, 318:407–412, 1994.

[69] T. Krick and L. M. Pardo. A Computational Method for Diophantine Approximation. In L. González-Vega and T. Recio, editors, *Proc. MEGA '94*, number 143 in Progress in Mathematics, pages 193–253. Birkhäuser, 1996.

[70] R.E. Ladner. On the structure of polynomial time reducibility. *J. ACM*, 22:155–171, 1975.

[71] J.C. Lagarias and A.M. Odlyzko. Effective versions of the Chebotarev density theorem. In *Algebraic Number Fields, Proc. 1975 Durham Symposium*, pages 409–464. Academic Press, N.Y., 1977.

[72] L. Landweber, R.J. Lipton, and E. Robertson. On the structure of sets in NP and other complexity classes. *Theoret. Comp. Sci.*, 15:181–200, 1981.

[73] S. Lang. *Algebra.* Addison-Wesley, second edition, 1984.

[74] L.A. Levin. Universal search problems. *Problems of Information Transmission*, 9:265–266, 1973.

[75] R.J. Lipton and L.J. Stockmeyer. Evaluation of polynomials with superpreconditioning. *J. Comp. Syst. Sci.*, 16:124–139, 1978.

[76] D.E. Littlewood. The construction of invariant matrices. *Proc. London Math. Soc. (2)*, 43:226–240, 1937.

[77] D.E. Littlewood. *The Theory of Group Characters and Matrix Representations of Groups.* Oxford University Press, 1940.

[78] G. Malajovich and K. Meer. On the structure of NP_C. *SIAM J. Comp.* To appear.

[79] D.A. Marcus. *Number Fields.* Springer Verlag, 1977.

[80] M. Marcus and H. Minc. On the relation between the determinant and the permanent. *Illinois J. Math.*, 5:376–381, 1961.

[81] D.K. Maslen and D.N. Rockmore. Generalized FFTs - a survey of some recent results. *DIMACS Series in Discrete Mathematics and Theoretical Computer Science*, 28:183–237, 1997.

[82] R. Merris. On vanishing decomposable symmetrized tensors. *Linear and Multilinear Algebra*, 5:79–86, 1977.

[83] R. Merris. Recent advances in symmetry classes of tensors. *Linear and Multilinear Algebra*, 7:317–328, 1979.

[84] R. Merris. Single-hook characters and Hamiltonian circuits. *Linear and Multilinear Algebra*, 14:21–35, 1983.

[85] M. Mignotte. Some useful bounds. In B. Buchberger, G.E. Collins, and R. Loos, editors, *Computer Algebra, Symbolic and Algebraic Computation*, pages 259–263. Springer Verlag, 1982.

[86] R. Motwani and P. Raghavan. *Randomized Algorithms.* Cambridge University Press, 1995.

[87] C.H. Papadimitriou. *Computational Complexity*. Addison-Wesley, 1994.

[88] C.H. Papadimitriou and K. Steiglitz. *Combinatorial Optimization: Algorithms and Complexity*. Prentice-Hall, 1982.

[89] C.H. Papadimitriou and S. Zachos. Two remarks on the power of counting. In *Proc. 6th GI conference in Theoretical Computer Science*, number 145 in LNCS, pages 269–276. Springer Verlag, 1983.

[90] N. Pippenger. On simultaneous resource bounds. In *Proc. 20th FOCS*, pages 307–311, 1979.

[91] J.S. Provan. The complexity of reliability computations in planar and acyclic graphs. *SIAM J. Comp.*, 15(3):694–702, 1986.

[92] J.S. Provan and M.O. Ball. The complexity of counting cuts and of computing the probability that a graph is connected. *SIAM J. Comp.*, 12(4):777–788, 1983.

[93] H.J. Ryser. *Combinatorial Mathematics*. Number 14 in Carus Math. Monographs. Math. Assoc. America, New York, 1963.

[94] A. Satyanarayana and A. Prabhakar. New topological formula and rapid algorithm for reliability analysis of complex networks. In *IEEE Trans. Reliability, R-27*, pages 82–100, 1978.

[95] C.P. Schnorr. Improved lower bounds on the number of multiplications/divisions which are necessary to evaluate polynomials. *Theoret. Comp. Sci.*, 7:251–261, 1978.

[96] U. Schöning. A uniform approach to obtain diagonal sets in complexity classes. *Theoret. Comp. Sci.*, 18:95–103, 1982.

[97] U. Schöning. Minimal pairs for P. *Theoret. Comp. Sci.*, 31:41–48, 1984.

[98] M. Shub and S. Smale. On the intractability of Hilbert's Nullstellensatz and an algebraic version of "NP $\neq$ P?". *Duke Math. J.*, 81:47–54, 1995.

[99] A. Sinclair. *Algorithms for Random Generation and Counting: A Markow Chain Approach*. Progress in Theoretical Computer Science. Birkhäuser, 1992.

[100] S. Skyum and L.G. Valiant. A complexity theory based on boolean algebra. *J. ACM*, 32(2):484–502, 1985.

[101] S. Smale. Complexity theory and numerical analysis. *Acta Numerica*, 6:523–551, 1997.

[102] S. Smale. Mathematical problems for the next century. *Math. Intelligencer*, 20(2):7–15, 1998.

[103] V. Strassen. Berechnung und Programm. I. *Act. Inf.*, 1:320–335, 1972.

[104] V. Strassen. Vermeidung von Divisionen. *Crelles J. Reine Angew. Math.*, 264:184–202, 1973.

[105] V. Strassen. Polynomials with rational coefficients which are hard to compute. *SIAM J. Comp.*, 3:128–149, 1974.

[106] V. Strassen. Algebraic complexity theory. In J. van Leeuwen, editor, *Handbook of Theoretical Computer Science*, volume A, chapter 11, pages 634–672. Elsevier Science Publishers B. V., Amsterdam, 1990.

[107] L.G. Valiant. Completeness classes in algebra. In *Proc. 11th ACM STOC*, pages 249–261, 1979.

[108] L.G. Valiant. The complexity of computing the permanent. *Theoret. Comp. Sci.*, 8:189–201, 1979.

[109] L.G. Valiant. The complexity of enumeration and reliability problems. *SIAM J. Comp.*, 8:410–421, 1979.

[110] L.G. Valiant. Reducibility by algebraic projections. In *Logic and Algorithmic: an International Symposium held in honor of Ernst Specker*, volume 30, pages 365–380. Monogr. No. 30 de l'Enseign. Math., 1982.

[111] L.G. Valiant. An algebraic approach to computational complexity. In *Proc. Int. Congress of Mathematicians*, volume 2, pages 1637–1644. Polish Scientific Publishers, Warsaw and Elsevier Science Publishers, Amsterdam, 1983.

[112] L.G. Valiant. Why is Boolean complexity theory difficult. In M.S. Paterson, editor, *Boolean Function Complexity: Selected papers from LMS Symposium, Durham, July 1990*, number 169 in London Mathematical Society, Lecture Notes Series, pages 84–94. Cambridge University Press, 1992.

[113] L.G. Valiant, S. Skyum, S. Berkowitz, and C. Rackoff. Fast parallel computation of polynomials using few processors. *SIAM J. Comp.*, 12(4):641–644, 1983.

[114] L.G. Valiant and V.V. Vazirani. NP is as easy as detecting unique solutions. *Theoret. Comp. Sci.*, 47:85–93, 1986.

[115] N.J. Vilenkin and A.U. Klimyk. *Representations of Lie groups and special functions, I, II, and III*. Number 72, 74, and 75 in Mathematics and Its Applications. Kluwer Academic Publishers, Dordrecht, 1991, 1992, and 1993.

[116] P.J. Weinberger. On Euclidean rings of algebraic integers. In *Analytic Number Theory, Proceedings of the Symposium in Pure Mathematics*, pages 321–332. Amer. Math. Soc., Providence, R.I., 1972.

[117] P.J. Weinberger. Finding the number of factors of a polynomial. *J. Algorithms*, 5:180–186, 1984.

[118] D.J.A. Welsh. *Complexity: Knots, Colourings and Counting.* Cambridge University Press, 1994.

[119] R.R. Willie. A theorem concerning cyclic directed graphs with applications to network reliability. *Networks*, 10:71–78, 1980.

List of Notation

$\mathbb{N}$	set of natural numbers
$\mathbb{Z}$	ring of integers
$\mathbb{Q}, \mathbb{R}, \mathbb{C}$	fields of rational, real, complex numbers
$\mathbb{F}_q$	finite field with q elements
$\operatorname{char} k$	characteristic of field k
$k[X_1, \ldots, X_n]$	polynomial ring in indeterminates $X_1, \ldots, X_n$ over field k
$\{0,1\}^*$	set of finite strings over alphabet $\{0,1\}$
S_n	symmetric group on n symbols
GL_m	group of invertible $m \times m$ matrices over $\mathbb{C}$
$\mathrm{U}(m)$	group of unitary $m \times m$ matrices
K_n	complete graph on n nodes
$K_{n,n}$	complete bipartite graph with two sets of nodes of size n
DK_n	complete digraph on n nodes
R_n	rectangular lattice on n^2 nodes
C_n	cubic lattice on $2n^2$ nodes
$R(\mu)$	rosette corresponding to positive integer μ
$\mathcal{E}$	graph property
$\mathrm{GF}(G, \mathcal{E})$	generating function corresponding to graph G and graph property $\mathcal{E}$
$\mathrm{PGF}(G, \mathcal{E})$	probability generating function corresponding to graph G and graph property $\mathcal{E}$
$L(f_1, \ldots, f_t)$	(total) complexity of $f_1, \ldots, f_t$
$L_{ns}(f_1, \ldots, f_t)$	multiplicative (or nonscalar) complexity of $f_1, \ldots, f_t$
$L^h(f_1, \ldots, f_t)$	oracle complexity of $f_1, \ldots, f_t$ with respect to h
$\mathcal{L}^h(f_1, \ldots, f_t)$	h-complexity of $f_1, \ldots, f_t$
$E(f)$	formula (or expression) size of f
$C(f)$	test complexity of f
$D(f)$	depth of f
$\tau(f)$	tau-complexity of integer polynomial f
Ω	quasi-ordered set
$f \leq g$	polynomial f is projection of polynomial g
$f \leq_p g$	family f is p-projection of family g
$f \leq_c g$	family f is c-reduction of family g
$\mathcal{D}_p$	poset of abstract p-degrees
$\mathcal{PD}_p$	poset of p-degrees of p-definable families
$\mathcal{PD}_c$	poset of c-degrees of p-definable families
$\lambda \vdash n$	partition of $n \in \mathbb{N}$
$\rho \models n$	partition of n in frequency notation
χ_λ	irreducible character of S_n belonging to partition λ

$\chi_{n,i}$	character belonging to hook partition $(n-i,1,\ldots,1)$
D_λ	irreducible representation of GL_m with highest weight λ
s_λ	number of standard tableaus on Young diagram of λ
d_λ	number of semistandard tableaus on Young diagram of λ
$G(\lambda)$	layered graph visualizing splitting behaviour of representations of GL_m
$\mathrm{mult}(\lambda)$	multiplicity of highest weight λ
$r(\lambda,\mu)$	number of vertical steps in skew hook for λ,μ
DET	determinant family
PER	permanent family
PER*	family of partial permanents
HC	family of Hamilton cycle polynomials
UHC	family of undirected Hamilton cycle polynomials
Cut^q	family of cut enumerators over $\mathbb{F}_q$
$\mathrm{HP}(G)$	generating function for s-t-Hamilton paths
$\mathrm{SW}(G)$	generating function for s-t-self avoiding walks
IM_λ	immanant polynomial corresponding to partition λ
$\mathrm{HI}_{n,i}$	hook immanant polynomial
CF_ρ	cycle format polynomial of ρ
$\mathrm{per}(A)$	permanent of matrix A
$\mathrm{Pf}(A)$	Pfaffian of matrix A
$\mathrm{hc}(A)$	Hamilton cycle value of matrix A
$\mathrm{im}_\lambda(A)$	immanant of matrix A corresponding to λ
$\mathrm{hi}_{n,i}$	value of hook immanant $\mathrm{HI}_{n,i}$ on matrix A
$\mathrm{cf}_\rho(A)$	cycle format value of matrix A
$\mathcal{CF}_\rho$	cycle format ρ property
$\mathcal{CL}$	clique property
$\mathcal{CO}$	connectivity property
$\mathcal{DI}$	dimer or perfect matching property
$\mathcal{FA}(F)$	F-factor graph property
$\mathcal{IS}$	Ising or closed graph property
$\mathcal{MD}$	monomer-dimer or partial matching property
$\mathcal{STC}$	s-t-connectivity property
Sat	satisfiability problem
#Sat	counting problem corresponding to Sat
Mod_pSat	mod $p=1$ decision problem corresponding to #Sat
#Cut	problem to count cuts of given weight in given graph
Mod_pCut	mod $p=1$ decision problem corresponding to #Cut
#Naesat	counting version of not-all-equal Sat problem
FNC^i	class of string functions computable by log-space uniform families of Boolean circuits of polynomial size and depth $O(\log^i n)$
FP	class of string functions computable by polynomial time Turing machines

$\mathrm{FP}^{\#\mathrm{P}}$	class of string functions computable in polynomial time using an oracle in $\#\mathrm{P}$
$\mathrm{Mod}_p\mathrm{NP}$	class of mod $p = 1$ decision problems corresponding to $\#\mathrm{P}$
NC^i	class of languages corresponding to FNC^i
NP	class of decision problems verifiable in polynomial time
NP_w	class in weak BSS-model corresponding to NP
P	class of languages decidable in polynomial time
PAR	decision problems decidable by parallel BSS-machines working in polynomial time using an exponential number of processors
PAR_w	class in weak BSS-model corresponding to PAR
PH	polynomial hierarchy
$\oplus\mathrm{P}$	parity polynomial time complexity class
PSPACE	polynomial space
P_w	class in weak BSS-model corresponding to P
$\#\mathrm{P}$	counting complexity class
$\#_p\mathrm{P}$	mod p counting complexity class
VNC^i	class of p-families computable by straight-line programs of polynomial size and depth $O(\log^i n)$
VNP	class of p-definable families
VNP^h	class of p-definable families relative to h
VP	class of p-computable families
VP^h	class of p-computable families relative to h
VQP	class of qp-computable families
VQP^i	set of p-families (f_n) satisfying $L(f_n) \le n^{O(\log^i n)}$
$\mathcal{C}/\mathrm{poly}$	nonuniform version of complexity class $\mathcal{C}$
$\mathrm{BP}(\mathcal{C})$	Boolean part of complexity class $\mathcal{C}$
$\pi(x)$	number of rational primes $\le x$
$\pi_K(x)$	number of primes of number field K with norm $\le x$
$\pi_S(x)$	number of primes $p \le x$ such that system (S) is solvable modulo p
$\pi_g(x)$	number of primes $p \le x$ such that integer polynomial g has a root modulo p
$P_\ell(x)$	Legendre polynomial of degree ℓ
$P_\ell^\mu(\cos\theta)$	associated Legendre function of order ℓ ($|\mu| \le \ell$)
$B_{i,j}(t)$	$m \times m$ matrix with entries 1 in diagonal, entry t at position (i,j) and zero elsewhere
$M(d)$	upper bound on complexity for symbolic multiplication of two polynomials of degree d
$\#\phi$	number of satisfying assignments of Boolean formula ϕ
$\mathrm{codim}\,Z$	codimension of affine or projective subvariety Z
$\deg Z$	degree of affine or projective subvariety Z
$i \le_\Gamma j$	there is a path from i to j in acyclic digraph associated with straight-line program Γ

$v(f)$	number of variables of polynomial f
$\mathrm{wt}(f)$	weight of polynomial f
$\mathrm{wt}(E')$	weight of set of edges E'
$\varphi(n) \gtrsim \psi(n)$	means $\liminf_{n \to \infty} \varphi(n)/\psi(n) \geq 1$
$C_n(h)$	n^{th} generic polynomial computed relative to h
$D_n(h)$	n^{th} generic polynomial defined relative to h

Index

Printing: Mercedes-Druck, Berlin
Binding: Stürtz AG, Würzburg